HMS WARRIOR

1860 to date

COVER IMAGE: **HMS** *Warrior*.
(www.billingboats.com)

First published in July 2017.

A catalogue record for this book is available from the British Library.

ISBN 978 1 78521 106 5

Library of Congress control no. 2016959368

Published by Haynes Publishing,
Sparkford, Yeovil,
Somerset BA22 7JJ, UK.
Tel: 01963 440635
Int. tel: +44 1963 440635
Website: www.haynes.com

Haynes North America Inc.,
859 Lawrence Drive, Newbury Park,
California 91320, USA.

Printed in Malaysia.

HMS WARRIOR

1860 to date

Owners' Workshop Manual

HMS WARRIOR
1860

An insight into owning, operating, maintaining and
restoring the Royal Navy's first armoured iron warship

Richard May

Contents

OPPOSITE HMS *Warrior* 1860. *(Mary416/Shutterstock)*

Acknowledgements

I have spent my working life in naval shipbuilding, having been involved with vessels as diverse as the most modern examples right through to square-rigged sailing ships. With an interest in maritime history, I found an immediate affinity when I became involved with HMS *Warrior*. She brings to life the huge leap forward that took place during the 19th century not only in Great Britain's industrial capability and technical innovations but also the social and human changes that were taking place at this time. All these aspects are reflected in this ship.

The restoration task was the largest and most challenging maritime rebuild ever undertaken and the result is an exhibit that is complete and tactile: you can sit on the mess deck where the crew sat and – quite literally – touch history. Biased though I am, *Warrior* is probably the best-presented historic ship in the world.

I have to thank the Warrior Preservation Trust for supporting the idea of this book, in particular the ship's Captain Tim Ash; Executive Officer Tony Ford; Campaign Director Fiona Postie; and all the staff and volunteers who have contributed both directly or indirectly in providing information.

Sincere thanks to Dr Louise Moon, Anna Bowman and Emily Pierce-Goldberg, the *Warrior* archivists, who have given their time and patience in finding information.

To Jonathan Falconer at Haynes Publishing I extend grateful thanks for providing guidance throughout this task.

Finally, to my wife and daughter who have tolerated me during the writing of this book. Many thanks.

I acknowledge the assistance given by the Warrior 1860 Preservation Trust (WPT); Gary Cook – artist; Kent Billing of Billing Boats, Viborg, Denmark, for kind permission to reproduce the image of his 1:100 scale model of *HMS Warrior* (visit www.billingboats.com for further information); Crossrail; Patsy Gaches; National Museum of the Royal Navy, Portsmouth; National Maritime Museum, Greenwich; National Portrait Gallery, London; Newham Archives and Local Studies Library; Museum of London; National Archives, Kew; Shutterstock; Walker Image Archive; and TopFoto.

There are a number of photographs in this book for which I have not been able to source the originator or copyright holder. If these should be discovered subsequent to publication then a suitable acknowledgement will be incorporated in any future edition.

Richard May MRINA
Portsmouth, February 2017

Introduction

The Historic Dockyard in Portsmouth provides the visitor with a wide spectrum of naval history that no other museum in the world can even begin to rival. One can see and move through the 16th-century Tudor times of the *Mary Rose* to the 18th century of Nelson's flagship *Victory*, then on to the 19th-century *Warrior* and onwards into the 20th century with the *M33* monitor ship from the Gallipoli campaign and the submarine HMS *Alliance* from the Second World War. Meanwhile, in the adjacent naval base, can be seen warships of the 21st century. Many of the dockyard buildings are historic in their own right and house museums and artefacts of our naval past.

Warrior is no exception. An hour on board will give a fast tour but two hours is better in order to see all the spectacular Victorian engineering that this ship has to offer. It also shows to great effect the huge transition that occurred in terms of Britain's industrial capability.

Warrior has been at Portsmouth since 1987 with over eight million visitors having come on board. She has a privileged location on her own jetty, and can be seen by visitors both outside and within the dockyard. The ship is the responsibility of the Warrior Preservation Trust, which is an independent charity that was established by the late Sir John Smith. The

BELOW HMS *Warrior*'s **arrival at Portsmouth in June 1987.**
(Warrior Preservation Trust – WPT)

ABOVE **HMS** *Warrior*, **set for a 21st-century dinner on board a 19th-century ship.** *(WPT)*

Trust has the enviable task of presenting the ship to the public with full-time crew, staff and volunteers. Education programmes are run, tailored for infants up to senior school pupils, while university and academic students are welcomed with deeper and more specialised presentations. Corporate hospitality, anniversary dinners, weddings, trade shows and other functions are also held on board. Wedding formalities are conducted in the captain's cabin and the reception held on the gun deck with guests seated at the mess tables as sailors did several generations earlier. These functions, together with the many films, documentaries and television programmes that have used *Warrior* as a location, provide the Trust with a significant financial income.

LEFT **A mess table prepared for a wedding breakfast.** *(WPT)*

As with all historic artefacts, preservation and finance are the two key aspects involved in keeping the ship in a presentable condition. The original iron ship itself gives little trouble when compared with the issues that arise from the significant amount of timber on the ship that is prone to deterioration.

The two major tasks undertaken since arrival in Portsmouth have been the replacement of the upper deck planking and that of the timber bulwarks. Despite the hours and dedication that the skilled restoration team in Hartlepool had given to restoring the deck with timber recovered from a demolished Bradford warehouse, as well as employing the traditional shipbuilders' methods of using oakum and Stockholm tar as the sealant between the planks, the timbers began to deteriorate after some twenty years exposed to the rigours of sun and rain. With the help of a Heritage Lottery Grant in 2001, the whole of the upper deck was stripped and relaid using teak and sealed with modern synthetic caulking (owing to its enormous advantages over the traditional

methods, it was used to ensure durability and water tightness). The ship's patron, HRH Princess Alexandra, ceremonially laid the final plank in February 2004.

Likewise, the timber bulwarks were found to be deteriorating and these too needed to be replaced. The dampness that occurs within the timber makes it vulnerable to a parasitic beetle known as the wharf borer (*Narcerdes*

FINAL PLANK LAID BY
HER ROYAL HIGHNESS
PRINCESS ALEXANDRA
4 FEBRUARY 2004

melanura). Though the beetle itself is harmless, the eggs that the female lays produce creamy white larvae that, despite their small size, are equipped with jaws that bore into wood and feed on the timber. They present a significant threat to timbers used in any marine environment. All the upper deck bulwarks are to be replaced and, as with all tasks, further work is sure to emerge.

Teak is no longer readily available in the quantities that would be needed to restore the entire bulwarks of the ship. The water bar is therefore in steel and the new timber is accoya, which is a chemically treated modern softwood that gives longevity equal to a hardwood but can be obtained from environmentally sustainable sources.

As part of the regular maintenance routine of the ship, approximately every ten years she is dry docked, primarily to remove the marine growth that is inevitable on any ship that does not go to sea, and then the underwater hull is inspected for any problems, followed by a new coat of paint. For a ship that is nearly 160 years old, her hull remains in remarkably good condition.

ABOVE Extensive deterioration is apparent in this view showing the timber below the water bar.
(Author)

RIGHT New timberwork around *Warrior*'s bowsprit.
(Author)

LEFT HMS *Warrior* in dry dock for her once-a-decade inspection. *(WPT)*

BELOW Portsmouth Historic Dockyard with HMS *Warrior* at her berth in the foreground, c.2010. *(Michael Stokes/ Shutterstock)*

Chapter One

The *Warrior* story

HMS *Warrior* was the largest and most powerful warship that the world had seen when she entered service in 1861. So powerful, indeed, that she never fired a shot in anger and can be thought of today as the nuclear deterrent of the Victorian age: nobody would have chosen to argue with *Warrior*. And she had a profound effect on naval architecture.

OPPOSITE HMS *Warrior*, the Royal Navy's first armoured iron warship, is moored permanently at Portsmouth. She is pictured from the Hard in February 2012. *(Steve Mann/Shutterstock)*

'A black vicious ugly customer as ever I saw, whale-like in size, and with as terrible a row of incisor teeth as ever closed on a French frigate.'
(Charles Dickens on seeing HMS *Warrior*)

In the beginning

Why was HMS *Warrior* built? To answer this question, first we need to understand the history and politics of the period. From 1799 until 1815 Great Britain was at war more or less continuously with Napoleon Bonaparte's France. Huge resources were committed to the war effort and the Royal Navy had the largest fleet of warships of any nation. But every ship relied on their sails and the prevailing winds.

In 1805 Horatio Nelson had shown Britain's dominance in naval power at the Battle of Trafalgar, decisively defeating the combined fleets of France and Spain. His flagship, HMS *Victory*, is preserved in Portsmouth Historic Dockyard a short distance from HMS *Warrior*.

In 1815 the Duke of Wellington brought the Napoleonic Wars to an end at the Battle of Waterloo. As a result, Napoleon was exiled to live his final years on the island of St Helena and Britain was at peace. Queen Victoria ascended to the throne in 1837 and Britain entered into the era known as *Pax Britannica*. Though the navy had reduced in size it was still the greatest and most powerful navy in the world and Britain still ruled the waves.

In 1848 Louis-Napoleon Bonaparte,

Napoleon Bonaparte's nephew, was democratically elected as the President of the Second French Republic. However, when he was blocked by the Constitution and Parliament from running for a second term, he organised a coup d'état in 1851, and then took the throne as Emperor Napoleon III in December 1852. During the first years of the new empire Napoleon's government imposed censorship and harsh repressive measures against his opponents. Some 6,000 people were imprisoned or sent to penal colonies. Thousands more went into voluntary exile abroad, many to Great Britain.

Napoleon III embarked on many great projects, among them the rebuilding of the Paris that we see today and, to the mounting concern of Britain, the expansion of the French armed forces.

France wanted to equal Britain's industrial capability and challenge her worldwide influence, so Napoleon set about doubling the size of the French Empire in Africa and Asia and to dominate Italy. However, Britain and France put aside their rivalry and were allies during the Crimean War against Russia 1853 to 1856.

Britain was impressed with the floating armoured batteries and guns that the French navy used against Russian fortifications. Not only did they destroy forts but they were also capable of receiving enemy shot without damage. A general impression was created in Britain that the Royal Navy was lagging behind in terms of such technological advances. Despite their co-operation in Crimea, Britain's relationship with France was at best tense, and when an Italian exile attempted to assassinate Napoleon III during his visit to London in 1858, it plummeted to new depths.

The Admiralty had always approached any technical innovations, and especially the idea of iron ships, with great caution and scepticism. They had built a variety of iron frigates during the 1840s but considered them unsuitable for combat. Many of these were paddle-steamers rather than screw-driven ships. Trials had demonstrated that gunfire would penetrate the iron hulls all too easily, shattering the iron into lethal fragments. Likewise paddle wheels were highly vulnerable to enemy gunfire. All these early vessels were either sold or converted

RIGHT
Louis-Napoleon III.
(Roger-Viollet/TopFoto)

into troopships, among them the ill-fated HMS *Birkenhead* that sank with heavy loss of life in 1852. Though the tragedy was caused by poor navigation, it turned the minds of the Admiralty even further against iron ships and towards the idea that wooden vessels would always prevail in the Royal Navy. Ships of the line that had been fitted with steam propulsion using propellers were not a success as the multitude of timbers from which they were constructed 'worked' when at sea and misalignment of the propeller shafts would occur. Steam engines were considered only as supplementary propulsion for use only when leaving and entering harbour, and in any case, steam boilers needed coal that was expensive – wind was free. The Admiralty saw no future in steam-driven iron battleships.

In 1858 France embarked on the construction of a powerful new ironclad class of warship, the first being named *La Gloire.* France was now challenging Britain's dominance of the seas.

'What a to-do and what to do'

Throughout 1858 the British newspapers became vitriolic in their condemnation of the government and the Admiralty for not responding to the French naval programme. Exaggerated claims of the capabilities of *La Gloire* and the number of armoured ships the French were building began to circulate. Public opinion was inflamed and a state of near hysteria pervaded Britain. In that same year the Liberal government of Lord Palmerston had fallen, to be replaced by Lord Derby's Tory administration.

On the naval issue, Lord Derby had to face the problem and find a resolution. The Admiralty itself was dominated by elderly ultra-conservative admirals with an average age of 74 who, in nearly all cases, had achieved their rank by political influence or patronage. Many of them had not even been to sea, or, if they had, it was a generation earlier. 'What was good enough for Nelson is still good enough today' was the general attitude. The very idea of the navy adopting armoured iron ships was viewed as nothing more than a passing fad. Furthermore, the ability of new guns to fire at greater ranges caused heated debate as many still believed that close combat and the boarding of an enemy ship was the predominant tactic of naval warfare.

Lord Derby came under pressure from no less a person than Queen Victoria, asking 'if the navy would be adequate in wartime'. A Parliamentary Committee of Enquiry known as the Derby Committee was formed. The Prince Consort Prince Albert took an interest in its formation, and wrote some very pointed and direct letters demanding that the committee should not be 'made up of old First Sea Lords that will stifle the enquiry. We need practical men who are not compromised.'

Among the significant 'practical men' were Sir John Pakington, appointed First Lord of the Admiralty and given the onerous task of having to decide what to do. In the meantime he had to hear all the arguments for and against, but was told to take particular notice of the views of Admiral Sir Baldwin Walker, Surveyor of the Navy. Walker was responsible for the warships built for the Royal Navy and had the advantage of being

RIGHT Admiral Sir Baldwin Wake Walker, Surveyor of the Navy (1848–61). He is pictured here while serving with the Turkish Navy in 1840. *(Public domain)*

FAR RIGHT Sir John Pakington (Lord Hampton). As First Sea Lord he commissioned Britain's first ironclad warship, HMS *Warrior*, in 1860. *(Public domain)*

held in high regard by the Lords of the Admiralty.

Despite being influenced by the construction of large iron commercial ships such as Brunel's *Great Britain* and the enormous *Great Eastern*, the largest ship in the world at the time, Walker still held reservations over whether iron warships would eclipse wooden warships as first-line battleships, especially based on the experiences with the earlier iron frigates. There was great support for the idea of adding armour plates to the sides of existing wooden warships and Walker did not object to this as it would be cheap to do. Pakington was sympathetic to Walker's views of lines of battleships carrying numerous guns engaging an enemy at close range, but he also accepted that well-armed frigates had a place in the order of battle. Despite all these conflicting arguments, he knew that something had to be done to meet the French challenge and the demands of his political masters.

Various proposals were tabled. The results of trials of new guns and armour plating were examined. Steam engines were evaluated. This resulted in Walker recommending to the Admiralty and the Derby Committee a proposal to build six ships, two with wooden hulls with ironclad armour, like *La Gloire*, and four iron ships. This six-ship proposal met immediate resistance from the Chancellor of the Exchequer Benjamin Disraeli, not just on cost that would have exceeded £1 million – a vast sum for that age –

but on political conflicts he had with both Sir John Pakington and Sir Baldwin Walker. If Disraeli had his way no new large warships would have been built. But Pakington, keenly aware that something had to be done, sought a solution.

Emerging from all this political manoeuvring, with what may be seen as a semi-compromise, was a proposal from Walker for the building of a single 36-gun screw frigate with an option to build a second vessel: iron being the preferred construction. The rationale behind this decision was for a powerfully armed and armoured frigate with good speed from steam engines that would be able to approach enemy ships quickly and deliver overwhelming destruction.

Decisions

Significant issues affected the decision making that led towards the final design. Timber was no longer readily available and what was available was being used to maintain the current fleet of wooden ships. New timber was reaching premium prices and having to be imported; irrespective of the passions held by the admirals for Britain's 'wooden walls', they were rapidly becoming an impossibility to build.

Steam-driven iron merchant ships had been at sea for many years – SS *Great Britain* first sailing in 1845 – and these ships were getting faster and bigger – in particular in terms of length. Yet since steam engines were not reliable or efficient by modern standards and relied on a supply of coal, every ship still had a full sailing rig. This was the balancing act that ships had to achieve in having good sailing capability and making the steam propulsion as effective as possible. In simple terms, wide ships are better under sail and narrow ships more efficient when steam propelled. Wooden warships were broad-beamed and relatively short, whereas ships such as the *Great Britain* and *Great Eastern* were by comparison long and thin. HMS *Victory*, for instance, has a length-to-breadth ratio (LBR) of 4, the ship being only four times longer than her width. In the final *Warrior* design the ratio is nearly 7, making her comparatively long and thin.

Despite the skills of the shipwrights of the period, long, thin ships in wood could not be built. Timber was not strong enough to withstand the longitudinal bending with the risk of a ship breaking her back. The largest wooden warship built up to that time was the 121-gun HMS *Victoria*, with both steam and sail, becoming the last three-deck wooden ship to enter service. However, she was heavily braced with large diagonal iron girders and her length to breadth ratio was the same as the *Victory*. As an aside, the very last wooden three-deck ship of the line to be constructed was HMS *Howe*, built in 1860 – the same year as *Warrior* – and though she ran trials under her steam engines, she never entered service, instead becoming an accommodation ship at Plymouth. Her only claim to fame was that, when broken up in 1921, a great deal of her timbers were used to renovate Liberty's department store on Regent Street, London, into the mock Tudor façade still seen today.

Walker was more than aware that the Naval Dockyards could only build wooden ships. They had very little experience of working with iron when compared with private shipbuilders. His requirement for the new frigate was intentionally biased for it to be an iron ship and for expediency he would have to invite private shipbuilders to respond with designs.

The original letter to the Board of the Admiralty clearly states Walker's requirements and is remarkable in the detail that he had brought together. The following is just an extract from a long specification.

Department of the Surveyor of the Navy

Confidential   **27th January 1858**

Frigate of 36 guns cased with Wrought Iron Plates – Design Obtain

With reference to the question of Building Ships to be cased in Iron to render them Shot-proof, I beg to state that having given this important subject my best consideration, and it appearing that the most judicial course would be not only to call on the Master Shipwrights in the Dockyards, but also to request some of the most eminent private shipbuilders who have had considerable experience in iron shipbuilding, to furnish designs. I beg to submit that the parties named in the margin be informed that their Lordships having under their consideration the subject of a shot-proof vessels would be glad to receive designs and suggestions for vessels of this description, and that the proposed particulars be sent for their information, observing that if they are disposed to furnish a Design not in accordance with these conditions, but which in their opinion would be better calculated to answer the intended purpose, their Lordships would be glad to receive it also.

As Iron appears to be the most suitable material for a ship of this kind both as regards strength and durability, the design should be for an Iron Ship, but if it is considered by any of the parties called on that a more satisfactory arrangement could be made with wood than iron, a plan and the particulars of a wooden ship may be forwarded for consideration; observing that in a wooden ship the armour plates must necessarily extend from stem to stern, whereas in an iron ship it might be considered advisable to limit their extent to about 200 feet of the middle points of the vessel, separating the part cased from the parts not cased by strong athwartship bulkheads, covered also with 4½ inch plates to extend down to about 5 feet below the plates on the side.

The ship to be masted and rigged as an 80 gun ship and to have sufficient steam power to give a speed of at least 13½ knots under steam alone when fully equipped with all stores onboard.

I further submit that the Private Shipbuilders be informed that it is important their Lordships should know the probable cost of such a vessel before coming to any decisions, that it is desirable that an estimate of cost of building her, and the time required, be furnished with the design, and that the information requested should be forwarded on the 1st next.

B W Walker Surveyor

The requirement was sent to 15 builders. It was the first time that private shipbuilders had been invited to submit their own designs and build a major warship, a monopoly that had hitherto lain with the Admiralty Dockyards.

At the same time Walker had asked Isaac Watts, the Chief Constructor of the Navy, to prepare a design. Watts was assisted by Chief Engineer Thomas Lloyd. An autocratic individual, Watts' experience came from designing wooden warships but it was recognised that he had considerable ability stemming from a long career in naval architecture.

All 15 shipbuilders responded to Walker's proposal, of which half the bidders were from Admiralty yards and three submitted designs for a wooden ship. The proposals were studied and all were rejected with doubts over the weight, the stability or whether the proposed engines had sufficient power.

Walker decided that his own design, prepared by Watts, was the best. Viewed through modern eyes this may seem somewhat nepotistic and self-serving, though at the time it was viewed as a brave decision. Walker was very aware that he needed to overcome the political hurdles, especially with Disraeli, and that control of the design details and costs were critical.

As First Lord of the Admiralty, Sir John Pakington approved the design and it was put before the Admiralty board in December 1858, where the proposal was endorsed. In turn, it received Cabinet approval in early January 1859 under an emergency programme to rebuild the navy. There were still reservations about building such a revolutionary warship and in a speech to parliament in February of that year Pakington said:

. . . the progress of the French ships has convinced the Admiralty that whatever the cost, we have no option, in the discharge of our duty but to commence the construction of iron cased ships and we have resolved that it is our duty to build at least two of these vessels.

Of the 15 shipbuilders who had responded, two were considered to have provided the most proficient replies, the first being Napier in Glasgow and the second Ditchburn and Mare on the Thames. The latter had considerable

experience of building iron ships and enjoyed an enviable reputation for quality. Charles Mare, who had run the company, found himself in financial difficulties around the time of the *Warrior* proposal, and his father-in-law Peter Rolt had taken over the business and re-formed it as the Thames Iron Works and Shipbuilding Company. Rolt was the Member of Parliament for Greenwich and Deputy Lieutenant for Middlesex and a consummate and successful businessman who would go on to make the Thames Iron Works the most prosperous yard on the river.

The order for *Warrior* was placed with the Thames Iron Works on 11 May 1859, having accepted their tender price of £190,225 and with the stipulation that the vessel would be launched within eleven months and completed three months later, excepting masts. Five days later the engine manufacturers John Penn & Son received the order to build the engines.

However, the politics continued, as Lord Derby's Tory government was replaced by a Liberal government in June 1859, again under the leadership of Lord Palmerston. Sir John Pakington was replaced by the Duke of Somerset and the Admiralty Lords were also changed. Though Palmerston believed in a large navy, both he and those he had appointed were sceptical of iron warships and they were not encouraged. But the press and public opinion was still running high, as if war with France was inevitable, and Palmerston had to be seen to be doing something. From the opposition benches of Parliament Sir John Pakington pressed Palmerston as to when the second iron ship would be ordered. Though Palmerston and his Admiralty Lords did not necessarily believe it was the correct thing to do, the order for the sister ship was placed with Napier of Glasgow in October 1859.

At this time both ships received their names: *Warrior* and *Invincible*. *Invincible*, however, had to be rapidly changed to *Black Prince* when it was discovered that the second French iron-clad had been given the same name.

Isaac Watts had taken Walker's fundamental requirement of a ship built around an armoured box. The box, or citadel as it became known, ran the 200ft (61m) length of the gun deck and down from the main deck to 6ft (1.8m) below the waterline. It not only protected the

	Guns	Crew	Length	Beam	LBR	Draught	Weight	Engines	Coal	Speed
Warrior	36	705	420ft	58ft	6.5	26ft	9,137t	5,267hp	850t	14kts
La Gloire	36	570	256ft	56ft	4.5	27ft	5,618t	2,500hp	625t	13kts
Victory	102	850	227ft	52ft	4	21ft	3,500t	none	–	8kts

armament but the engines, boilers, secondary steering and the areas where the majority of the crew worked. This was the strength of the ship and in simplistic terms the bows and the stern were added at either end of the box.

It can be seen that Watts' final design incorporated a number of constructional techniques which were more in line with those used on wooden ships, and it could be construed that the ship was overdesigned, resulting in a heavy but extremely durable hull. In all respects, however, the design was sound.

The table above shows that the ship had the following particulars and the comparison with *La Gloire* and *Victory* shows how revolutionary *Warrior* was in size and power:

The build

The Thames Iron Works was located at Bow Creek. It stood at the mouth of the River Lea, which enters the Thames at Blackwall just to the west of the Royal Victoria Docks and where the modern Excel Exhibition Centre now stands. The works immediately started to construct a new slipway as *Warrior* was three times heavier than any vessel they had previously built. Having received the order on 11 May 1859 it was only 14 days later, 25 May, that they laid the keel. Work progressed apace.

Issues began to emerge almost immediately, and one that caused an early hiatus was the idea of introducing the ram bow on to *Warrior* as there were those in the Admiralty who thought this a significant weapon for steam-driven vessels. Last used by the Greek galleys many centuries before, its value was questionable in the age of guns. But it became a heated point of debate between advocates and opponents who wanted a typical clipper bow. Watts had to compromise and *Warrior*'s bow was modified during the build to a heavy reinforced stem post that would withstand an impact but retain the clipper bow shape. This additional structure added weight to the bows of the ship that was later to cause problems.

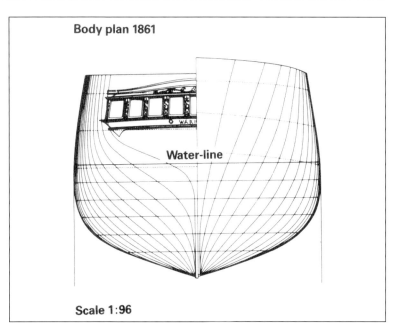

Body plan 1861

Water-line

Scale 1:96

ABOVE Body plan. *(Gary Cook)*   **BELOW** Half-frame. *(Gary Cook)*

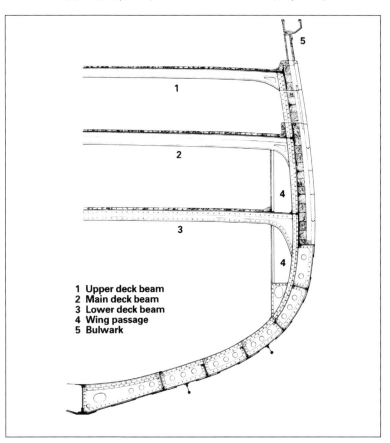

1 Upper deck beam
2 Main deck beam
3 Lower deck beam
4 Wing passage
5 Bulwark

RIGHT **Thames Iron Works.** *(Newham Archives and Local Studies Library)*

BELOW **An 1850s map of the River Thames showing Bow Creek and the Thames Iron Works.** *(Public domain)*

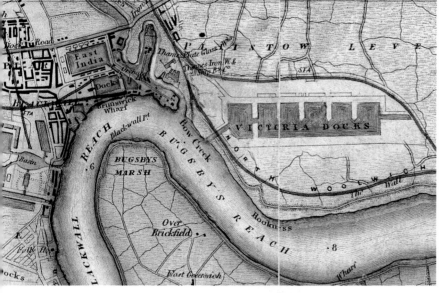

RIGHT **The Butterley Company's name moulded on one of** *Warrior*'s **beams, which they struggled to supply in time and in sufficient quantity. They are known as 'Butterley Bulbs' and were among the earliest examples of extruded iron beams.** *(Author)*

The Thames Iron Works quickly discovered they had undertaken a significant build, not only suffering from design changes imposed upon them but from the supply of materials – especially the iron beams from the Butterley Company of Derbyshire, who struggled with both delivery and quality. This resulted in a fractious correspondence between the two companies.

The delivery date was not going to be met. In October 1859 the Admiralty wrote to Peter Rolt of the Thames Iron Works requesting '. . . greater exertion to be used in proceeding with the work'. The company replied '. . . in the consequences of her novel construction they have not been enabled to progress so rapidly as they could have wished, but hence forward they will cause work to progress to their Lordships' satisfaction'.

But the builder was having difficulties. The weather that winter was exceptionally harsh, the build slipping further behind programme. They needed more labour and, worse still, were running out of money as they had received only half of the first-stage payment. In February of 1860 the Admiralty released further funds that enabled the Thames Iron Works to employ more labour. Dependent on skills, the wages ranged from 3s to 7s 6d for a six-day 54-hour week.

By now the ship was rising on the banks of Bow Creek to dominate the skyline. She was attracting ordinary sightseers, dignitaries and

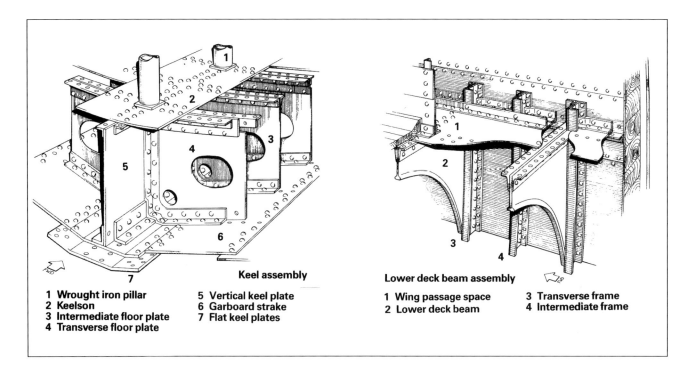

Keel assembly

1	**Wrought iron pillar**	**5**	**Vertical keel plate**
2	**Keelson**	**6**	**Garboard strake**
3	**Intermediate floor plate**	**7**	**Flat keel plates**
4	**Transverse floor plate**		

Lower deck beam assembly

1	**Wing passage space**	**3**	**Transverse frame**
2	**Lower deck beam**	**4**	**Intermediate frame**

royalty alike. Prince Alfred (Queen Victoria's second son), the Lords of the Admiralty, ambassadors and naval attachés from numerous countries all came to see progress being made.

Warrior was breaking new ground and old traditions in many spheres of the navy. Among them was the decision to paint the ship black. She would look impressive with her white sails spread and yellow funnels. As a consequence, all ships that followed her were painted black

ABOVE Keel assembly and lower deck beam assembly. *(Gary Cook)*

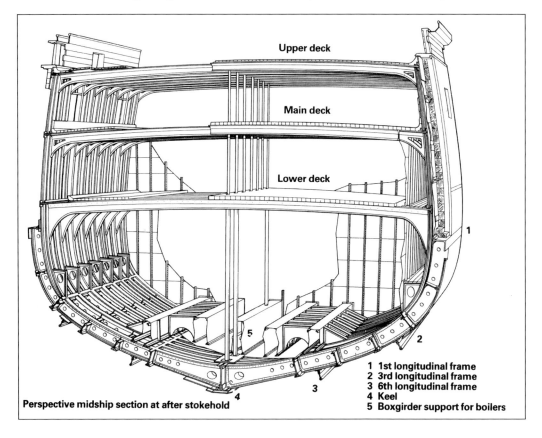

Perspective midship section at after stokehold

1	**1st longitudinal frame**
2	**3rd longitudinal frame**
3	**6th longitudinal frame**
4	**Keel**
5	**Boxgirder support for boilers**

LEFT Perspective midship section at after stoke hold. *(Gary Cook)*

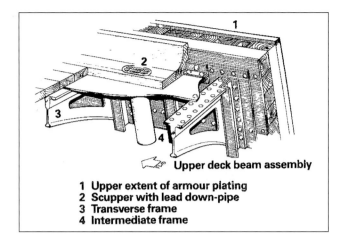

Upper deck beam assembly

1 Upper extent of armour plating
2 Scupper with lead down-pipe
3 Transverse frame
4 Intermediate frame

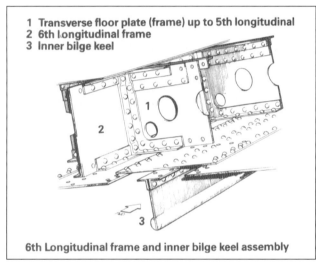

1 Transverse floor plate (frame) up to 5th longitudinal
2 6th longitudinal frame
3 Inner bilge keel

6th Longitudinal frame and inner bilge keel assembly

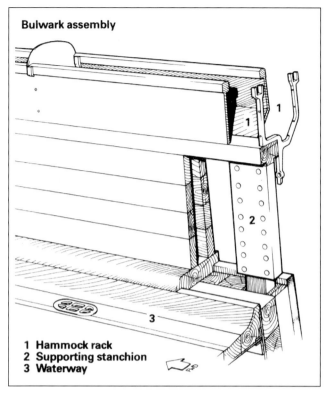

Bulwark assembly

1 Hammock rack
2 Supporting stanchion
3 Waterway

ABOVE LEFT Upper deck beam assembly. *(Gary Cook)*

ABOVE Bulwark assembly. *(Gary Cook)*

LEFT Sixth longitudinal frame and inner bilge keel assembly. *(Gary Cook)*

RIGHT The launch ways at the Thames Iron Works at Bow Creek uncovered in 2012 during the Crossrail construction programme. *(Crossrail)*

and the Royal Navy became known as the 'Black Battle Fleet'. Indeed, Emperor Napoleon III once said of *Warrior* that she was 'a black snake amongst the rabbits'.

Taken from Captain John Wells' book *The Immortal Warrior* is a wonderful extract from a reporter who was writing for the magazine *Temple Bar*. His description brings alive the activity witnessed on a summer's day when he visited the shipyard:

Waiting to be ferried across the creek the Warrior was before us, lying in a side cut on the opposite bank. Her huge hull of iron, painted red and covered in scaffoldings, rose up as a five or six storied house . . . workmen are seen clinging on to her in almost every spot; some are carpenters . . . but most are iron workers and the din of hammers is so loud that we are obliged to shout as if in a gale of wind. We creep along the scaffolding . . . scared by the shout 'below there' when down comes a rivet. But the men go on quite unconcerned, many of them seated all day on a wet plank close to the mud, pulling at the handle of a drill with the same kind of action that a rower uses; in this way the iron is pierced for the bolts. . . .

Of the stern, he noted:

. . . looking upwards at this stupendous piece of forge work, is the stern post, which goes right away to the upper deck . . . it took months of ceaseless forging to get this gigantic limb of Warrior into shape.

The discourse continued:

. . . entering through a gunport without stooping . . . heavy blows resound on every side and it requires rather a sharp lookout to avoid being hit as you pass by some stout hammerer . . . hoarse cries and angry shouts from the men are answered by shrill cries from the boys who, armed with long pincers, rush madly by with red hot bolts and take flying leaps . . . like so many young imps. Now the ship appears to be on fire in 50 places and this gives the whole scene a strange

BIRD'S-EYE VIEW OF THE THAMES IRONWORKS SHIPBUILDING YARD, SHOWING TWO WAR-SHIPS UNDER CONSTRUCTION.

ABOVE Thames Iron Works and Shipbuilding Company in a photograph dated 1902. Extensive work is still being undertaken. *(Newham Archives and Local Studies Library)*

BELOW Construction of an iron cruiser at the Thames Iron Works. This may be HMS *Minotaur* that was ordered immediately after *Warrior*. The photograph is indicative of the construction of such a vessel and the thousands of men who were employed. *(Newham Archives and Local Studies Library)*

character of wild and imposing fierceness and power . . . nothing short of the terrible experiences of war would call up such tremendous effort.

Progress was being made but correspondence still flowed between the Admiralty and the Thames Iron Works over dates for launching and completion. The 11 months from build to launch was not going to be achieved, and money was again a significant factor. The Admiralty accused the Thames Iron Works of not employing enough workers and they

retaliated by pointing out that they had been subjected to many design changes to the ship during the build.

The humiliation for the country, the government, the Admiralty and those working on *Warrior* was acute when it was learned that the French ironclad *La Gloire* had been completed by August 1860.

Further money was found and by the time of her launch £250,000 had been spent. At last a launch day was agreed for 29 December 1860: six months later than anticipated. The date was dictated by the spring tide due on that day as high water was needed to launch this huge ship. Dredging work was undertaken to ensure sufficient water.

Before the launch the English winter turned cold and there was snow. The temperature dropped each day and became one of the coldest winters ever recorded. The Thames began to freeze over.

The launch

It could have been expected that the honour of launching the most powerful warship Britain had ever built would fall to royalty, but the Admiralty Board, after due deliberation, turned to the previous First Sea Lord, Sir John Pakington, to perform the ceremony (although Pakington received the invitation only ten days before the event). Politics was the most probable motive behind the invitation. As the largest warship ever built, the *Warrior* could also be said to have been the largest risk the Admiralty had ever embarked upon; that being the case, the man who had authorised her construction ought to take the responsibility if she was a failure.

On 29 December 1860 crowds began to gather that grew to thousands as the launch time approached. The ship had been fitted with three temporary masts flying the Admiralty flag, the Royal Standard and the Union Flag. A launch platform had been erected at the head of the building berth with *Warrior*'s bows and figurehead rising above it. A band played.

The dignitaries attending were the Admiralty Board, Members of Parliament from both the Commons and the Lords, ambassadors, naval

attachés and the directors of the Thames Iron Works, many of whom brought their wives.

The freezing conditions had caused contingencies to be put in place, not least hundreds of men on the upper deck, hydraulic rams on the launchways and paddle tugs with towing hawsers attached to the stern.

At 2.30pm, Sir John Pakington and guests assembled on the platform. Unlike modern launches, there was no religious ceremony. With these simple words Sir John christened the ship: 'I name this ship *Warrior*. May God bless her and all who sail in her.' But the ship did not move.

Immediately the hydraulic rams pressed hard against the launchways. The hundreds of men on the upper deck moved in a co-ordinated mass from one side to the other to induce *Warrior* to unstick from the frozen ways. The paddle tugs began to pull, but it was a full 20 minutes before she moved. Once she did, though, there was no stopping her. Sir John Pakington had only moments to break a bottle of wine on her bows and wish 'God speed the *Warrior*.' The band struck up 'Rule Britannia' and the ship slid into the waters of the River Lea.

Fitting out

Though afloat, *Warrior* was empty as very little had been installed. The ship was towed into the Royal Victoria Docks for fitting out. The tasks involved included installation of her engines, boilers, masts and rigging. Partly because of the cold weather progress with the external work was slow. This delay assisted Penn, however, as they were manufacturing the most powerful steam engines ever built and they too were experiencing delay.

The engines were assembled piece by piece in the factory and tested. Ten boilers were constructed, being simple box boilers. In March 1861 the first parts of the engines and boilers were brought by barge from Deptford to the ship and installed.

ABOVE Ready for launching, December 1860. *(WPT)*

BELOW The launch. Though photography was in its infancy at the time, no photographs of the launch or of the dignitaries who were present are known to exist. *(WPT)*

Once again increasingly acrimonious correspondence passed between the Admiralty and Thames Iron Works. The arguments were the same: slow progress, many design changes being accommodated and more money being needed.

At the same time as the engines and boilers were being installed, Chatham Dockyard delivered the masts. A barge fitted with a shear legs crane towing the timber masts and bowsprit behind came alongside *Warrior*. The mainmast was 120ft tall and 40in in diameter. As soon as the lower masts were in place with their stays and shrouds, the topmasts were installed, followed by the yards. The Chatham riggers worked in appalling weather. By May 1861 *Warrior* was starting to look complete.

The first of the crew began to join the ship from June and the slow transition from a building site to an active ship began. Changes to the ship were occurring right to the very end of her build. It was discovered that the French *La Gloire* had been fitted with a tower on her main deck for the command's protection when in action and the Admiralty demanded the same for *Warrior*. When installed it became known as the 'Rifle Tower'. Though the original intention was to construct it of timber, the captain designate Arthur Cochrane suggested it was armoured with rifle loopholes around the periphery. Access was from below via a hatch.

Having installed it late in July 1861 there was an immediate view that it was of little use. If a compass was taken into the tower, the iron seriously affected its accuracy. The loopholes were so small that visibility was negligible and the practicality of firing a rifle through them bordered on the ridiculous.

On 8 August 1861 Captain Arthur Cochrane came aboard with due ceremony and took command. The ship's flags were hoisted and most significantly the commissioning pennant was flown. HMS *Warrior* was a ship of the Royal Navy and was destined to change the navies of the world.

It had taken 19 months to complete her, some 6 months late, and she had cost £377,292 to build, no less than £180,000 over budget. This figure would have been enough to construct two wooden first-rate ships. But there was no comparison between *Warrior* and her predecessors.

Hull	£282,284
Machinery	£74,409
Mast and yards	£3,756
Rigging/boats	£15,952
Engineers' stores	£891
Total	£377,292
2017 inflation at 110%	£41,502,120

Sir Baldwin Walker, who had been such a driving force in creating *Warrior*, had been placed in an increasingly invidious position. His role had been subjected to the political changes of naval policy by a government that was being influenced by diverse and various sources of what the future navy should be. His advice was no longer of significance to the Admiralty Board. He had been forced to have Isaac Watts design a smaller version of *Warrior* and four ships were ordered: *Defence, Resistance, Hector* and *Valiant*. They were expensive to build, performed poorly and were judged to have only a quarter of the power of *Warrior*. Walker had never agreed with building these ships and shortly before *Warrior* was completed he resigned. It was a sad end for a man who brought visionary foresight to the future Royal Navy.

Warrior's statistics

Length overall	418ft	127.4m
Waterline length between perpendiculars	380ft	115.8m
Beam	58ft	17.7m
Draught	26ft	7.9m
Tonnage	9,210 tons	9,358 tonnes
Height of foremast	169ft	51.5m
Height of mainmast	175ft	53.3m
Height of mizzenmast	138ft	42.0m
Sail area plain sail	37,546sq ft	3,436m^2
Trysails and stunsails	18,000sq ft	1,672m^2
Maximum speed – steam	14.3kts	26.5kph

In fairness to the Admiralty, they had recognised that *Warrior* and *Black Prince* were not enough on their own to ensure Britain's domination of the oceans and so further significant ships were ordered, the first being another 'Warrior' class, but one armoured from bow to stern. This was HMS *Achilles*, which was

LEFT HMS *Achilles*, considered the third of the 'Warrior' class, was ordered in 1861 and entered service in 1863. *(National Museum of the Royal Navy – NMRN)*

ordered from Chatham Dockyard. It signalled the start of the huge building programme of over 40 iron ships that encompassed the fast-evolving technology of the day.

Trials

Warrior was complete. She was now to embark on a career that was not spectacular – she did not fight any battles – but she was to change the world as she became the catalyst for taking naval warships from the era of the wooden naval ships that had existed for centuries to the Dreadnoughts of the 20th century. She was the datum from which future warships were measured.

Warrior left Victoria Dock on 8 August 1861 and made her way under steam to Greenhithe where she anchored. Large crowds of sightseers, both ashore and in boats, came to see her – many were even allowed on board. Among these visitors was Charles Dickens.

It had been intended to install the guns when she arrived at Portsmouth but the Admiralty were keen to see her fully equipped as soon as possible. This being the case, orders were given that the armament be sent from Woolwich Arsenal by barge and shipped on board. At this time only 68lb cannon were installed.

It took time to ship the guns and time for the crew to establish their routines and master the basics of operating the ship. It was not until September that *Warrior* made passage for her home port of Portsmouth. The voyage served as the first of a long series of trials that would find out exactly what she could do.

Though in an overall sense *Warrior* performed well on this first voyage, a number of

issues began to emerge that were inevitable for such a revolutionary design. Of significance was the tendency to plunge into the sea, bringing water over the bows and on to the main deck. The legacy of the heavy reinforced bow as the appeasement to the advocates of ramming was causing *Warrior* to trim down by the bows. The other, and possibly more significant, problem was that the steering at best was poor and at worst did not answer the helm at all.

The gunports could not be made watertight when closed and sea water entered the gun deck. Among other issues, there were concerns about marine growth 'fouling' the ship's bottom – fairly extensive growth had developed during the fitting-out period when the ship lay dormant in Victoria Dock.

On arrival in Portsmouth *Warrior* was dry-docked to remove the appendages of the launching ways that were still attached to the ship. On examination it was found that the

BELOW HMS *Warrior* at Plymouth in what is possibly the first photograph of the ship in 1861. *(National Trust)*

'fouling' was not as bad as first seen as the
voyage from the Thames had worn away most
of the growth. The Admiralty chemist W.J. Hay
had been an early pioneer of anti-fouling paint
and one of his schemes, known as 'Hay's
Composition', was applied. Despite this, *Warrior*
was docked at least every year of her first two
commissions to ensure she had a clean hull and
maintain her speed against coal burned.

Number 10 dry dock in Portsmouth was at this
time the only dry dock in the country large enough
to accommodate ships of the size of *Warrior*. While
she was in Portsmouth, the Prime Minister Lord
Palmerston went to see her. Other notable visitors
included the Duke of Cambridge, the Bishop of
Winchester, Grand Duke Constantine and a host

of other dignitaries. Though welcomed and treated
with respect, these visits did not assist in bringing
Warrior to operational readiness.

Palmerston rather begrudgingly acknowledged
that she was a powerful ship but thought *La
Gloire* was superior as she was armoured from
bow to stern, whereas *Warrior's* bows and stern
would be vulnerable. 'Paste board ends' was his
description. Palmerston was certainly no expert
on warship construction and his comments may
have been naïve, but they led to him supporting
the building of even more powerful warships.

In October 1861 she put to sea for trials of
her capabilities to establish precisely what her
performance was. To the west of Portsmouth in
Stokes Bay was the measured mile with a
pair of large masts set on the shore exactly
1 nautical mile apart. As the ship steamed past
the first mast, stopwatches were set running,
then stopped when she passed the second
mast: the calculation established the knots per
hour speed of the ship. *Warrior* did six runs, three
in either direction, both with and against the tide.

Warrior, 9,400 tons of iron ship, achieved an
unprecedented speed of 14.3kts under steam
alone. There was not the slightest vibration. This
was a unique achievement as all previous steam-
driven ships, especially the wooden warships,
were notorious for vibrating at full speed.

A conversation was alleged to have taken
place between Sir John Pakington and Peter
Rolt, the Thames Iron Works chairman after the
speed trials. Pakington mused: 'I often wonder
how I mustered sufficient courage to order the
construction of such a novel vessel.' Rolt's
supposed reply was: 'I often wonder how I
mustered sufficient courage to undertake its
construction.'

This conversation may be apocryphal but the
performance of *Warrior* would have brought a
sense of huge relief and pride to both of these
men. From Rolt's perspective, she transformed
his business. He received orders for further
iron warships not only from the Admiralty but
from overseas navies, making the Thames Iron
Works a hugely successful company.

In October *Warrior's* sister ship *Black Prince*
arrived. But Napier, her builders, had not fully
completed her and she had been jury-rigged
and brought to Portsmouth for completion.

In November *Warrior* undertook a 24-hour full

power trial using both steam and sail to establish coal consumption and open-sea performance. She averaged 16kts with the wind and just under 10kts against it. During these trials, gunnery exercises were conducted, the overall results of which were deemed 'satisfactory'. In November she took passage from Plymouth to Portsmouth, completing the voyage in 10 hours. Running against the tide, she achieved a maximum speed of 17.5kts under combined steam and sail. This proved to be an unprecedented speed that was not matched for nearly 10 years.

At this time *Warrior* came the closest to military action of her career. Though she was always destined to be part of the Channel Squadron facing the French naval threat, events further abroad began to affect British thinking. No less a person than Prince Albert the Prince Consort was keenly aware of the developing situation in Mexico where a civil war was raging. He wrote to Lord Palmerston recommending that Britain should send *Warrior* as part of a proposed Franco-Spanish expeditionary force. Palmerston replied that it would not be good to show the French the qualities of *Warrior*, which they would no doubt copy. With great credit to Prince Albert, he agreed with the Prime Minister and withdrew his suggestion. In fact Prince Albert never actually saw *Warrior* as he died in December 1861.

At the same time there were rumours and speculation circulating in the press that Britain would intervene in the American Civil War that had begun in 1861. Britain remained largely neutral, though with a bias towards the Confederacy. There was no plan for *Warrior* to get involved, but the Royal Navy kept a keen eye on developments on that side of the Atlantic all the same.

In December *Warrior* berthed in Portsmouth Harbour to have a number of improvements made and to rectify some of the problems that had become apparent during her trials phase. The most significant was the removal of eight 68lb cannon from the gun deck to be replaced with the same number of Armstrong's 110lb rifled guns. The Captain, Arthur Cochrane, requested two further 110lb guns for the upper deck as 'chase guns'.

To rectify the earlier issues, the gunports were modified using vulcanised rubber seals to make

them watertight. Attention was also paid to the steering gear, which received a new, heavier tiller. Cochrane was perceptive in recognising that *Warrior*'s steering problem was the rudder itself that was far too small for a ship of her extreme length and weight. Despite his request for a larger rudder, the Controller of the Navy did nothing other than prevaricate with vague replies. *Warrior*'s steering was to haunt her all her life.

In January 1862 *Warrior* sailed from Portsmouth on what was called 'the second experimental cruise' to Lisbon via Plymouth. On leaving Plymouth she sailed into a force 11 gale. Sails were split, spars cracked, one of the ship's boats was nearly lost and *Warrior*'s bow dug into the seas bringing huge amounts of water on to the upper deck.

Despite this, *Warrior* arrived at Lisbon and Cochrane wrote that the ship '. . . did not strain and any of the armour loosen . . .' This test brought further acclaim that *Warrior* was the strongest and most capable ship in the navy.

The British press were for once generous in their praise of *Warrior* and did not hold back in rounding on those within Parliament, the Admiralty and others who had been, and remained, critical of the decision to build this revolutionary ship.

It was accepted by all that Britain had, with *Warrior* and *Black Prince*, the two most powerful warships ever to take to the seas.

Before we follow the career of *Warrior* with the Royal Navy, we take a look at the construction and layout of the ship in Chapter 2.

Chapter Two

Anatomy of *Warrior*

HMS *Warrior*'s sheer size is but the first of the many surprising and innovative things that visitors will see as they tour the five decks of the most powerful warship of its era. This chapter provides an overview from stem to stern, from quarter deck to engine room, of HMS *Warrior*.

OPPOSITE *Warrior*'s 5,400ihp Penn trunk engine.

Introduction

A sailor arriving on HMS *Warrior* from a navy that had only 'wooden walls' similar to HMS *Victory* would have been taken aback by the sheer size and space of this huge new ship. There were unheard of facilities like unlimited supplies of fresh water, bathrooms for all the crew, a laundry and clothes drying room, and engines that powered the ship irrespective of wind and tide. *Warrior*'s layout of five decks shows the balance that was required to fight a war, but also to facilitate the daily lives of her crew and their needs.

The upper deck

T he gangways bring visitors on to the upper deck. This open deck is dominated by the three masts, their spars and the mass of rigging. The extensive deck planking is of teak wood. There are no sea views from *Warrior* as the bulwarks are intentionally high, primarily to act as breakwaters in reducing 'wetting' of the deck in heavy seas and as protection of the crew when in action. The exception is at the stern where visitors can stand on the raised platform with its panoramic view of Portsmouth Harbour.

Along the top of the bulwarks are the hammock stowages where the crewmen placed their folded hammocks. At the front of the ship is the bowsprit which, when extended, could reach 70ft (21m) over the ocean. In the adjacent bulwarks are openings that led to the 'heads', the ship's latrines that are the gratings either side of the bowsprit.

Predominant on the fore deck is one of the 110lb Armstrong rifled guns, known as a chase gun. On the deck the brass races and pivot

CENTRE *Warrior*'s stern bulwark and ensign staff. (Author)

LEFT The hammock stowages along the top of the bulwarks. In the event of the ship foundering, the hammocks would float off and serve as rudimentary lifebuoys. (Author)

RIGHT The heads (toilets). Being at the head of the ship gave the toilets their name. Despite being the most technologically advanced warship in the world, the crew's heads were little changed from earlier ships. Cubicle toilets were built inside the bulwarks shortly after entering service.
(Author)

plates allow the gun to be moved into four different positions.

Stowed inside the bulwark are two double-coned buoys painted red and green. These served as markers that could be attached to the anchor by a lanyard to show where the anchor was as the ship swung on the tide.

Moving aft, sheep and cattle pens were found against the bulwarks. The two funnels rise above the deck and between them is the rope walkway that is the forward bridge. Far removed from a modern ship's bridge, it was from here that the captain and navigator, exposed to the winds and elements, would command the ship when under steam.

Stowed either side of the after funnel are five ship's boats. Only the two launches can be

BELOW Overhead ash rail. The ash bucket was lifted from the stoke hold and then pulled across the upper deck to the hatch in the bulwark and the contents dumped overboard.
(Author)

BELOW Funnels and cowls: the funnels are for smoke from the boilers, the cowls for drawing air into the boilers. The black funnel is the galley exhaust. *(Author)*

1 Stern.
2 Mizzen royal.
3 Armstrong 110lb chaser gun.
4 Spanker.
5 Mizzen top gallant.
6 Mizzen top.
7 Mizzen mast, 138ft.
8 Main mast, 175ft.
9 Main topsail.
10 Main top gallant.
11 Main royal.
12 Cutter gig.
13 Air intake cowls.
14 Galley exhaust.
15 Fore mast, 169ft.
16 Fore topsail.
17 Fore top gallant.
18 Fore royal.
19 Skylight gratings.
20 Hammock stowages.
21 Armstrong 110lb chaser gun.
22 Crew heads.
23 Bow sprit.
24 Figurehead.
25 Reinforced stem.
26 Bow.
27 Keel.
28 Forward bridge.
29 Funnels.
30 Launches.
31 Rifle tower.
32 Gig.
33 Aft bridge.
34 Propeller.
35 Rudder.

RIGHT The painting scow was used for cleaning and painting the waterline of the ship. This example was recovered from the moat of one of the Palmerston Forts at Portsmouth and could be contemporary with *Warrior*. *(Author)*

RIGHT The two launches also had smaller boats stowed inside. One launch was reportedly used as a chicken coop. *(Author)*

FAR RIGHT The mainmast rises 175ft (53.3m) above the deck. *(Author)*

seen as the smaller boats were stowed within the launches to save space.

Along the length of the deck are skylight gratings opening on to the lower decks. At sea they would be sheeted over with canvas covers, but in fine weather they were opened up. These gratings could be lifted out to permit the supply of materials and stores into the ship.

On the restored *Warrior*, glazed window frames have been fitted. Though these are not of the period, they keep the ship dry.

Further aft is the upper deck capstan used for a multitude of tasks including raising or deploying the anchors or lifting the numerous ship's boats into the water.

The after rope bridge is where the command would be when under sail or combined steam and sail as it provided the better view of the

RIGHT Aft bridge. Unlike modern ships' bridges it was an exposed position. The rifle tower can be seen below. *(Author)*

FAR RIGHT Binnacle compass on the aft bridge. *(Author)*

set of the sails. Meanwhile, the deck aft of the bridge was known as the quarter deck. Traditionally this was a raised deck from the era of wooden ships but on *Warrior* it is simply a continuation of the main deck; however, it was recognised as an area where only the captain, officers and duty crew were permitted.

Below the aft bridge is the armoured rifle tower. Copied from the French *La Gloire*, it was of questionable use and never used as such. Aft of the rifle tower and just before the mizzenmast is the upper deck set of steering wheels known as the 'cruising wheels'. Outboard are the davits for two cutters, the gig and galley.

At the stern is another Armstrong 110lb gun and deck races. This gun could be manoeuvred into five separate positions. On the deck within the gun races is a large brass-edged circle with perforated brass plates set into it. This is the portable deck panel above the propeller well that was lifted out when raising the propeller.

The jolly boat sits on davits at the very stern of the ship.

The main deck

There are seven ladders from the upper deck down to the main deck. When setting or taking in sail, large numbers of seamen would need to get aloft as fast as possible and so all the ladders would be used. The original ladders

LEFT Copied from that on the French *La Gloire*, the rifle tower proved of little value as visibility from it was poor and it very rapidly became redundant. *(Author)*

LEFT The aft gun races: the 110lb gun could be moved to five different positions on deck. The large brass circle is the lift-out panel above the propeller lifting well. *(Author)*

were iron-runged and a number survive deep within the ship.

Going down the forward ladder takes visitors on to the main deck where the forward end is known as the cable deck. Here the space is dominated by the chain cables that are attached to the anchors and made off (secured) around the bitts (large black cylindrical bollards). Right forward, adjacent to the bowsprit where it

BELOW LEFT One of the seven ladders down to the lower deck. *(Author)*

BELOW One of the ship's original iron ladders deep in *Warrior*. *(Author)*

RIGHT The manger: others would have been built on the upper deck, depending on how long the ship was to be away from harbour (and therefore from fresh supplies). *(Author)*

FAR RIGHT Warrant and petty officers lived on the cable deck (pictured here). *Warrior* is afloat and the anchor cables are doing their job in securing the ship. *(Author)*

passes out of the ship, is a temporary manger for livestock – complete with sheep.

The foremast can be seen where it passes down through the decks. Only two guns occupy this space. When built there were additional guns but these were moved aft to alleviate *Warrior*'s bow-down problem.

The cable deck was the home for the warrant officers and petty officers and is where

they would sling their hammocks and eat with the luxury of their own cook and cubicle heads.

Moving aft through the armoured bulkhead with its heavy doors and into the armoured citadel, or box, is the gun deck which is dominated by the rows of guns. Each side within the citadel was occupied by 13 guns. Between each gun sits a mess table, benches and shelves. In the deck head can be seen

RIGHT Warrant officers' cubicle heads. *(Author)*

FAR RIGHT Armoured door into the armoured citadel. *(Author)*

RIGHT Portside of the cable deck. The space is now used for lectures, demonstrations and social functions. *(Author)*

ABOVE Hammocks rigged on the gun deck where 650 men slept – where there was a hook there was a hammock. In the background can be seen other hammocks 'made up' and ready to be stowed on the top of the bulwarks. *(Author)*

RIGHT The gun deck – *Warrior*'s 'teeth'. *(WPT)*

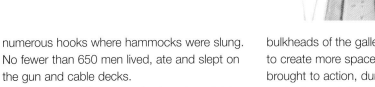

numerous hooks where hammocks were slung. No fewer than 650 men lived, ate and slept on the gun and cable decks.

The depth of the gunports show how thick the ship's side plating is, due to the armouring and its teak wood backing bolted to the iron ship. The gunport covers would normally have been closed and this deck would have been dark with only candles for illumination.

In the centre, the large iron structures enclose the forward and aft boiler uptakes, between which the galley is located. The

bulkheads of the galley could be folded inwards to create more space when the guns were brought to action, during coaling or when anchoring evolutions were being undertaken.

The large black and white painted capstan halfway along the deck is known as a Brown's patent steam-driven capstan and this replaced the original manual capstan during the 1875 refit.

Throughout this deck can be seen black circular hatches under the mess tables and guns; these are the coaling scuttles where embarked coal was shovelled down into the bunkers.

BELOW LEFT
Gunports were only opened for gunnery practice and the gun deck was a poorly lit area even in daytime. At night only candles and oil lamps provided illumination. *(Author)*

BELOW The ship's galley. *(Author)*

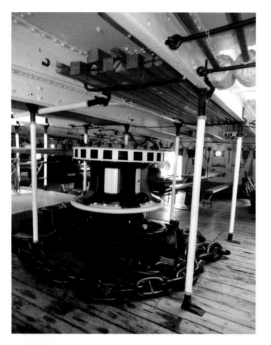

Towards the aft end of the gun deck is the second capstan that can be connected to the upper deck capstan. To work each manual capstan required 88 men, so when both were operated, 176 men were needed.

Passing through the after armoured bulkhead on to what was known as the half deck, we come to the area where the senior officers of *Warrior* lived. When anchoring astern, further bitts (bollards) were used, the chains of which passed through the captain's cabin.

Throughout the ship are racks of rifles, pistols and cutlasses. The half deck accommodated significant numbers. These weapons were intended to be used more for arming shore parties than repelling boarders.

Six guns were located on the half deck. Nowadays four guns are folded back to provide space for social functions and weddings. The

ABOVE The master's desk. The compass on the deck was taken aloft on the wooden foremast to obtain a 'true' reading as the other compasses were influenced by *Warrior*'s iron structure. *(Author)*

ABOVE The commander's cabin. *(Author)*

LEFT The commander had a bunk bed (or possibly a cot bed) and a cannon. His cabin is now used by the Portsmouth Registrar where couples who have been married on *Warrior* sign the register. *(Author)*

gun that was once placed in the commander's cabin has been removed.

The three cabins on the half deck accommodated the master, who was the ship's navigator and the commander, who ran the ship, while the most impressive cabin was

LEFT The captain's cabin. *(Author)*

RIGHT The sickbay. *(Author)*

BELOW Drunkenness, leave-breaking and selling clothes were the most common reasons for time spent in the cells located on the cell flat. *(Author)*

reserved for the captain, who had his sleeping cabin to the starboard side.

Behind the captain's cabin was the space for the tiller and the propeller-raising well as well as – most importantly – the captain's toilet.

The lower deck

The lower deck is the most intense in the diversity of stores, offices and facilities that kept the ship running and operational. Divided into watertight sections, the most forward compartment is the gunners' store, accessible only from the main deck. This area is now used as a paint store and is not open to visitors. The next compartment is the sickbay, which is available for escorted viewing.

Once again a ladder runs down from the main deck into the warrant officers' flat. This is the space where the gunner, boatswain, carpenter and engineers were all accommodated. The foremast passes down through this area. Nowadays this is not open to visitors, instead being used as the mess space for the *Warrior*'s current crew and volunteers.

Cell flat

The next ladder down from the main deck takes visitors to the cell flat that by its very name is where the ship's prison is on the port

ABOVE **The sail room.** *(Author)*

side. Drunkenness and insubordination were the principal reasons for earning a few days alone here. Opposite, on the starboard side, is the sail room.

On the cell flat a doorway has been cut through the armoured bulkhead. The armour and teak backing on the bulkhead can be clearly seen. This leads to the forward handing room where the gunpowder charges were brought from the magazine below and up to the gun deck.

The next space was the engineers' flat where maintenance and repairs would have been carried out. This now houses the Warrior Preservation Trust's workshop.

Leading on from this is the bathroom flat. There were two bathrooms, one on either side of the boiler casing. Only the port one is accessible and contains not only the bathrooms but also the laundry with the space dominated by the washing machines.

RIGHT **Carbolic soap was available for use in the bathtubs.** *(Author)*

BELOW **Wash tubs where dirty clothes were soaked.** *(Author)*

LEFT **The access door cut into the armoured bulkhead.** *(Author)*

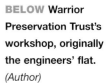

BELOW **Warrior Preservation Trust's workshop, originally the engineers' flat.** *(Author)*

ABOVE Washing machines with their mangles mounted above. *(Author)*

RIGHT One of the remotely operated sliding watertight doors installed during the 1864–67 refit. *(Author)*

BELOW Crew's kitbag racks on the seaman's flat. *(Author)*

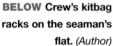

Dirty clothing was rinsed in the wooden buckets against the ship's side and then placed into the washing machines. These were heated by a steam hose brought from the stoke hold and then turned manually with the large handle at the end. Mounted on top are the mangles, again turned by hand, that squeezed the water from the clean laundry. The deck of the bathroom/laundry was covered in either lead or zinc sheeting and the water from the baths, washing machines and buckets was tipped on to the floor and allowed to drain down into the coal bunkers to prevent coal dust explosion and methane gas developing. Stokers and engineers had to bathe after each watch, boys once a day; officers had their own individual bathtubs in their cabins.

In the forward corner of the bathroom is the spare blade for the propeller. The bulkhead into the next compartment has a mechanically driven watertight door that could be closed remotely from the deck above. Passing through this door and into a space dominated by the uptake casing from the aft stoke hold, on either side of the casing are the numerous bag racks where the crew would have stowed their kitbags.

The two large white wheels are Downton pumps that were used to provide water to a fire main and to pump out ballast water. If the 850 tons of coal was used up, serious stability issues would have occurred and so ballast water was allowed to flood the double bottom.

BELOW A Downton pump for discharging ballast water and firefighting. *(Author)*

When coal was embarked the water would be pumped out of the ship, retaining stability.

The next compartment is the cable locker flat and is a large space. At the far end is the mainmast passing through and adjacent is the top of the cable lockers with the anchor cable passing down to the lockers below. On each side of the flat are large sail bins for the spare 'suits' of sail. Arranged in the centre is a display of rum casks, tubs, measuring jugs and barricoes from which the sailors' rum tot was issued. This is directly above the spirit room.

An open glazed hatch allows a view down into the shell room where the 110lb explosive ammunition was stored for the Armstrong guns.

The large black painted gear wheel is the drive mechanism from the 1875 Brown's capstan on the gun deck. There would have been a small steam 'donkey' engine adjacent that drove the gear wheel.

Passing through another sliding

RIGHT **The cable lockers. The mainmast is behind.** *(Author)*

BELOW **The starboard sail locker.** *(Author)*

BELOW RIGHT **Rum: one tot in the morning for the crew . . . well watered down with three parts water.** *(Author)*

ABOVE LEFT The shell room for the 110lb Armstrong gun's explosive ammunition. *(Author)*

ABOVE The gear wheel for the Brown's capstan. A small steam donkey engine drove the gear wheel but it has not been rebuilt for *Warrior* as this capstan was not in the ship in 1860. *(Author)*

LEFT Royal Marines' flat with bag racks. The space is open to the engine room that is below the iron grillage. *(Author)*

watertight door into the Marines' flat, which is above the engine room, there are two further Downton pumps, one of which can still be operated. The bag racks here were exclusively for the Royal Marines. Also present in this area is a lathe and grinding wheel used by the engineers.

The open grillage in the centre of the space sits above the engine room; the engine can be seen operating from here. The adjacent offices are for the medical dispensary, band instruments and the Royal Marines' slop clothing. Meanwhile, through a further door, we move on to the lower steering flat where the third set of steering wheels are situated, and to port is the issuing room from where each mess

LEFT The third set of wheels for use in an emergency. *(Author)*

RIGHT **In the issuing room the mess cooks collected the food allowance for the day.** *(Author)*

cook collected the daily rations. Opposite is the crew's slop clothing store.

There were two handing rooms on this deck that were above the magazine but only one survives on the starboard side. The doorway through to the gunroom was cut during *Warrior*'s time as a depot ship after her active service had finished. The original solid bulkhead would previously have served as the social divide between officers and the crew. The gunroom would have accommodated the midshipmen, who were young men and boys under training to become officers. Around the space are examples of midshipmen's chests, each containing his personal effects and the tools of his selected skill. Over to starboard is the cabin for the assistant surgeon, the clerk's office and a pantry.

Moving through to the wardroom, the dining

RIGHT **The gunroom, home of the midshipmen and the chaplain.** *(Author)*

BELOW **Midshipmen's chests.** *(Author)*

BELOW RIGHT **The wardroom was the home for 17 officers. Here it is set for 'harbour routine'. When at sea all crockery and the table were cleared away and the officers ate in their cabins.** *(Author)*

RIGHT **In the issuing room the mess cooks collected the food allowance for the day.** *(Author)*

RIGHT The mizzenmast terminated in the wardroom. *(Author)*

BELOW A typical officer's cabin was often very cluttered with personal possessions. *(Author)*

BELOW RIGHT At the very bottom of the hull is the bilge, some 30ft (10m) below the waterline. *(Author)*

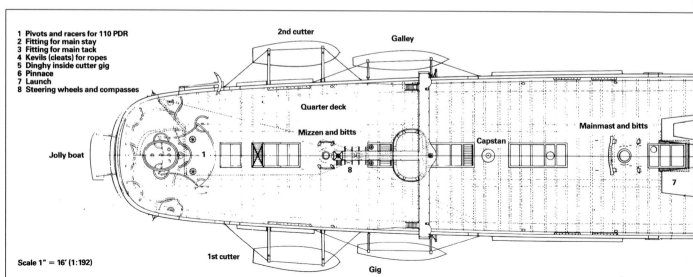

1 Pivots and racers for 110 PDR
2 Fitting for main stay
3 Fitting for main tack
4 Kevils (cleats) for ropes
5 Dinghy inside cutter gig
6 Pinnace
7 Launch
8 Steering wheels and compasses

2nd cutter

Galley

Quarter deck

Mizzen and bitts

Mainmast and bitts

Capstan

Jolly boat

1

8

7

Scale 1" = 16' (1:192)

1st cutter

Gig

table is set for dinner. The dinner service on board has been reproduced by Royal Worcester who made the ship's original set and each piece carries the WR inscription which stands for 'wardroom' and not *Warrior*. By tradition a photograph of Queen Victoria and her consort Prince Albert are displayed.

The mizzenmast terminates in the wardroom, and around the periphery of the space are cabins for the captain of Marines, first, second, third, fourth and fifth lieutenants, the paymaster and the chief engineer.

Further cabins are situated through the door of the wardroom partition and these now accommodate the Warrior Preservation Trust administration offices. They would originally have been cabins for the first and second

Marines officers, the chaplain, assistant surgeon, assistant chief engineer and the surgeon himself.

The hold

This is the lowest of the five decks in the ship and is situated some 30ft below the deep waterline. The forward space was the forward magazine but has not been restored, instead being converted into the access way for visitors to the stoke hold. Passing through the impressive stoke hold with its boilers, visitors come to the bottom of the mainmast, behind which is the ladder up to the engine room. From there it's past the bread store and then up the ladder back to the orlop deck.

ABOVE LEFT The stoke hold with its ten boilers. *(Author)*

ABOVE The engine room. *(Author)*

BELOW AND OVERLEAF All deck plans by Gary Cook.

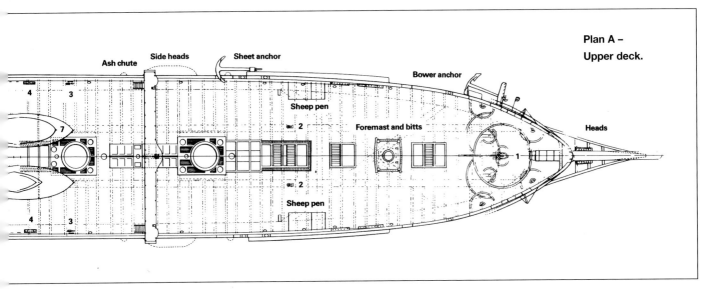

Plan A –
Upper deck.

Ash chute   Side heads   Sheet anchor   Bower anchor

Sheep pen   Foremast and bitts   Heads

Sheep pen

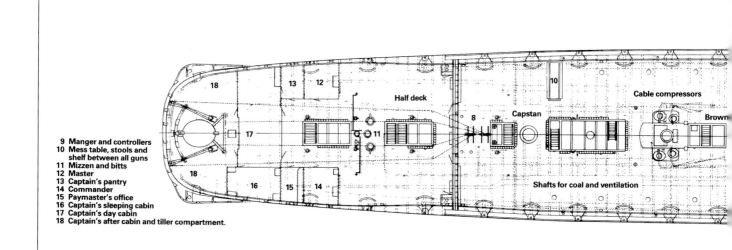

9 Manger and controllers
10 Mess table, stools and shelf between all guns
11 Mizzen and bitts
12 Master
13 Captain's pantry
14 Commander
15 Paymaster's office
16 Captain's sleeping cabin
17 Captain's day cabin
18 Captain's after cabin and tiller compartment.

Half deck

Capstan

Cable compressors

Brown

Shafts for coal and ventilation

1 Wardroom pantry
2 2nd Marine officer
3 Asst Surgeon
4 1st Marine Officer
5 Asst Chief Engineer

6 Chaplain
7 Surgeon
8 Captain of Marines
9 1st Lieutenant
10 5th Lieutenant

11 2nd Lieutenant
12 3rd Lieutenant
13 4th Lieutenant
14 Paymaster
15 Chief Engineer

16 Asst Surgeon
17 Clerks office
18 Pantry
19 Handing scuttles
20 Magazine ventilator

21 Issuing room
22 Slop room
23 Dispensary
24 Marines' slops
25 Chronometers and band in

Scale 1″ = 16′ (1:192)

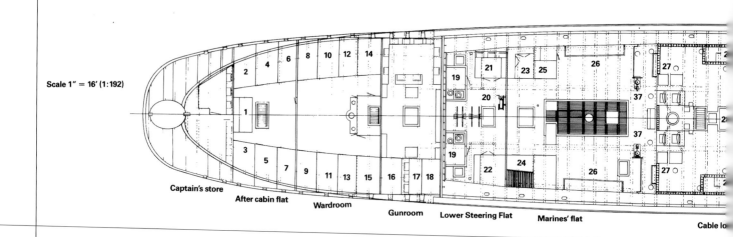

Captain's store   After cabin flat   Wardroom   Gunroom   Lower Steering Flat   Marines' flat   Cable lo

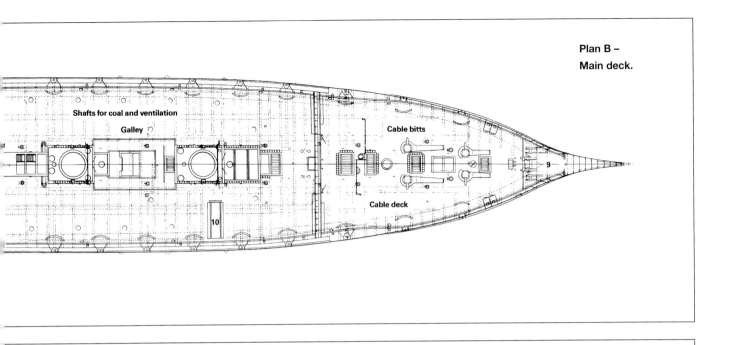

Plan B –
Main deck.

Shafts for coal and ventilation

Galley

Cable bitts

Cable deck

9

10

Plan C –
Lower deck.

26 Bag racks
27 to Shot and cable lockers
28 to Spirit room
29 Sail bin
30 Bathroom

31 Cells
32 Sail room
33 Boatswain
34 Carpenter
35 Gunner

36 Engineers' mess
37 Downton pump

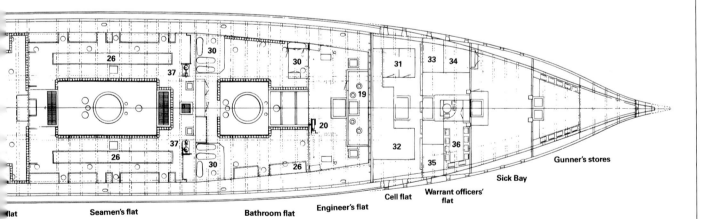

26

30

30

31

33   34

37

19

20

32

36

37

35

26

30

26

Gunner's stores

Sick Bay

Warrant officers'
flat

Cell flat

Engineer's flat

Bathroom flat

Seamen's flat

flat

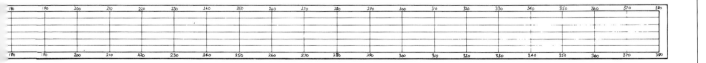

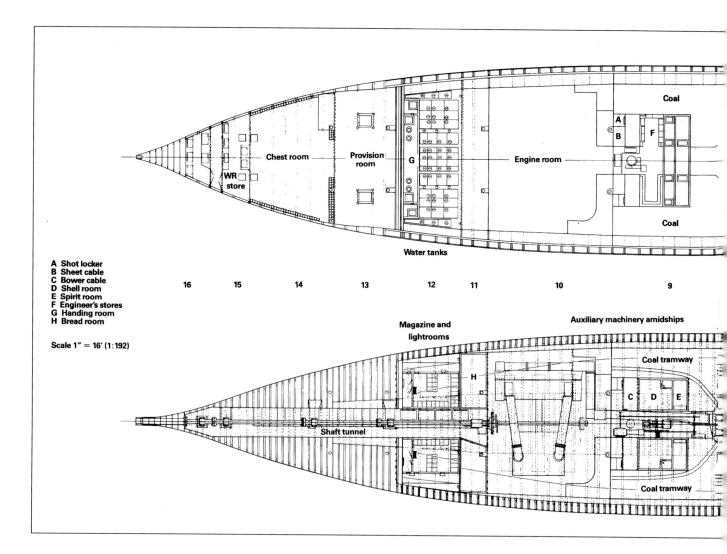

A Shot locker
B Sheet cable
C Bower cable
D Shell room
E Spirit room
F Engineer's stores
G Handing room
H Bread room

Scale 1" = 16' (1:192)

Chest room

WR store

Provision room

G

Engine room

Coal

A
B
F

Coal

Water tanks

16   15   14   13   12   11   10   9

Magazine and lightrooms

Auxiliary machinery amidships

Coal tramway

H

Shaft tunnel

C D E

Coal tramway

Sails and masts

Conscious of her length, there had at one time been serious consideration given to installing four masts – indeed plans had been prepared. However *Warrior* reverted to a standard rig as for an 80-gun wooden ship of the line, comprising three square-rigged masts. These were the foremast, mainmast and mizzenmast, and though strictly not a mast, one has to include the bowsprit.

As with everything in the Royal Navy, standardisation was the immovable element of Admiralty policy. *Warrior* had to fit into one of the categories for her sailing rig and having been deemed to be an 80-gun ship, the 'rigging warrant' would have already been written. This listed the size of masts and spars, the size and number of sails, the quantity of rope, blocks and every component that such a ship would

need and provided the authority to the riggers and boatswain to draw those items from the stores. If it was not in the warrant, you did not get it. Ships' rigging warrants have been issued since before Tudor times and still exist to this day on the most modern warships in the navy.

Though *Warrior* and her engines were built by commercial businesses, the rigging of a warship remained exclusively with the Royal Dockyards and Chatham Dockyard had the task of supplying and installing her rig.

Despite Isaac Watts' thoughts on iron masts, traditional wooden ones were installed. Made up from baulks of pinewood glued and pinned together, iron hoops were then forced down the mast (which was slightly tapered) to clamp the timbers. Each mast is made of three separate sections; the lower mast passes down through the decks on to the keelson at the very bottom of the ship; the exception is the

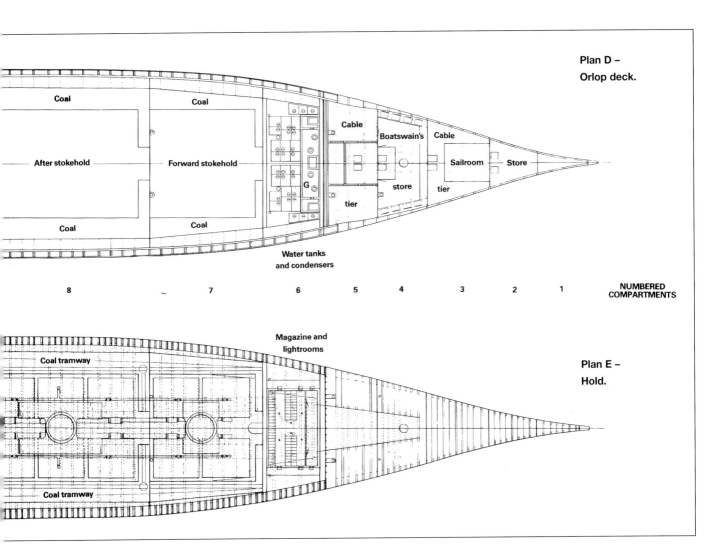

Plan D – Orlop deck.

Coal

After stokehold

Coal

Coal

Forward stokehold

Coal

G

Cable tier

Boatswain's store

Cable tier

Sailroom

Store

Water tanks and condensers

| 8 | | 7 | 6 | 5 | 4 | 3 | 2 | 1 | NUMBERED COMPARTMENTS |

Coal tramway

Magazine and lightrooms

Plan E – Hold.

Coal tramway

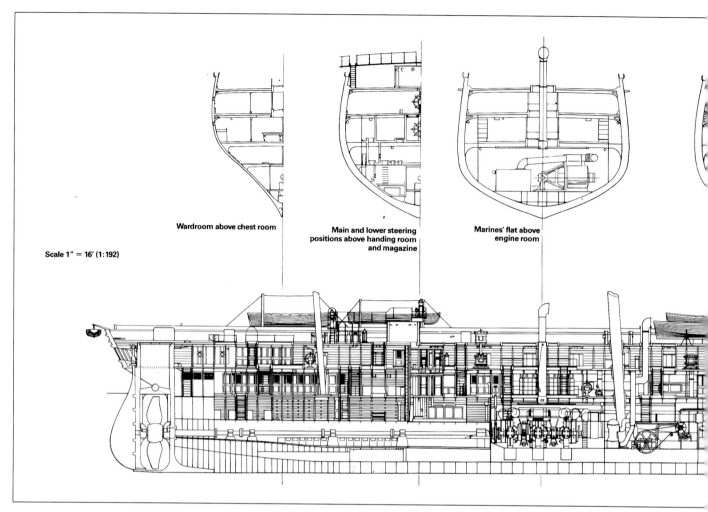

Scale 1" = 16' (1:192)

Wardroom above chest room

Main and lower steering positions above handing room and magazine

Marines' flat above engine room

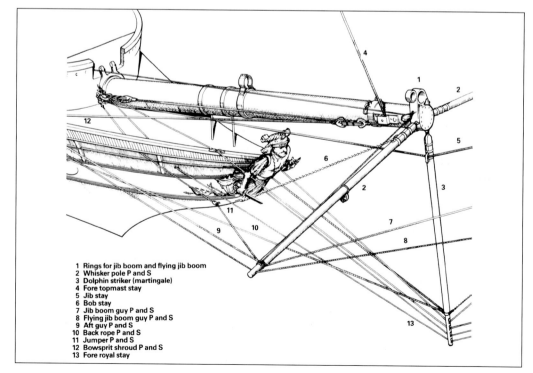

RIGHT Bowsprit.

(Gary Cook)

1 Rings for jib boom and flying jib boom
2 Whisker pole P and S
3 Dolphin striker (martingale)
4 Fore topmast stay
5 Jib stay
6 Bob stay
7 Jib boom guy P and S
8 Flying jib boom guy P and S
9 Aft guy P and S
10 Back rope P and S
11 Jumper P and S
12 Bowsprit shroud P and S
13 Fore royal stay

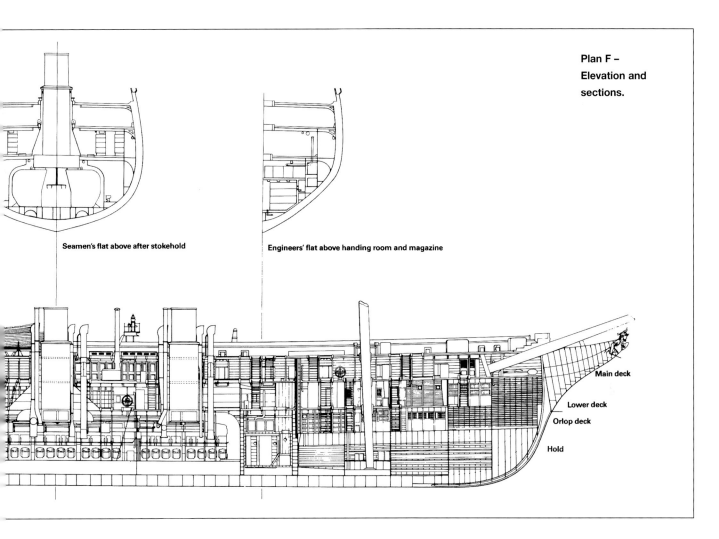

Seamen's flat above after stokehold

Engineers' flat above handing room and magazine

Main deck

Lower deck

Orlop deck

Hold

mizzenmast (the after mast) which terminates
in the wardroom due to the propeller shaft
tunnel passing underneath. The upper mast is
then secured to the lower mast. The topmast
is secured to the upper mast. On *Warrior* the
masts are not evenly spaced. The foremast is
positioned well forward, but the mainmast is
well aft and closer to the mizzenmast.

The heights of the masts were determined
by the width of the ship, so that the shrouds
and ratlines had the greatest spread in providing
support. Built into the masts and the bowsprit
was a copper lightning conductor.

The bowsprit is comprised of three sections.
The fixed part passes down into the ship and
terminates on the cable deck. On to this is fixed
the moveable jib boom that can be extended
and then to this is attached the flying jib boom.

When first built, the bowsprit was 49ft long,
the jib also 49ft and the flying jib 52ft. However,
following the trials, and to alleviate the ship's

bows-down condition, the bowsprit was cut
down to 25ft (with a smaller diameter) with a
42ft jib boom and a 45ft flying jib.

The vast array of ropes and wires around

BELOW Mile and miles
of rope. *(Author)*

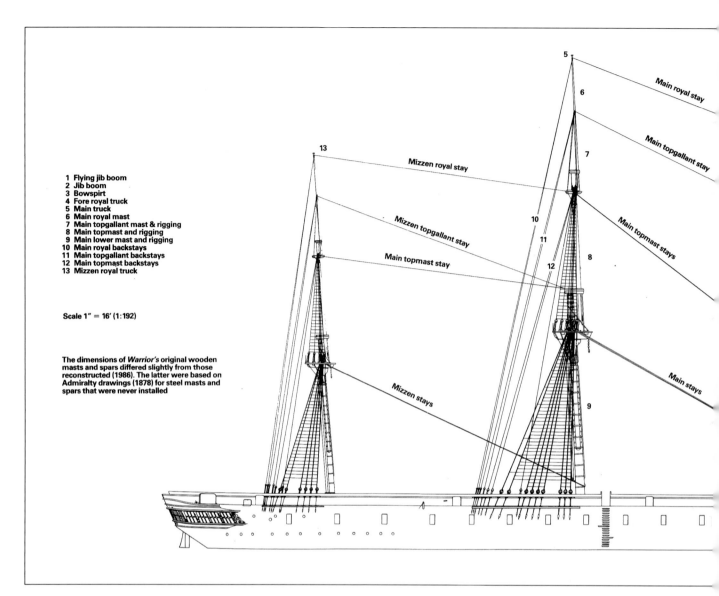

1 Flying jib boom
2 Jib boom
3 Bowsprit
4 Fore royal truck
5 Main truck
6 Main royal mast
7 Main topgallant mast & rigging
8 Main topmast and rigging
9 Main lower mast and rigging
10 Main royal backstays
11 Main topgallant backstays
12 Main topmast backstays
13 Mizzen royal truck

Scale 1″ = 16′ (1:192)

The dimensions of *Warrior's* original wooden masts and spars differed slightly from those reconstructed (1986). The latter were based on Admiralty drawings (1878) for steel masts and spars that were never installed

Main royal stay

Main topgallant stay

Mizzen royal stay

Main topmast stays

Mizzen topgallant stay

Main topmast stay

Main stays

Mizzen stays

ABOVE Standing rigging. *(Gary Cook)*

RIGHT The stays that support the masts. *(Author)*

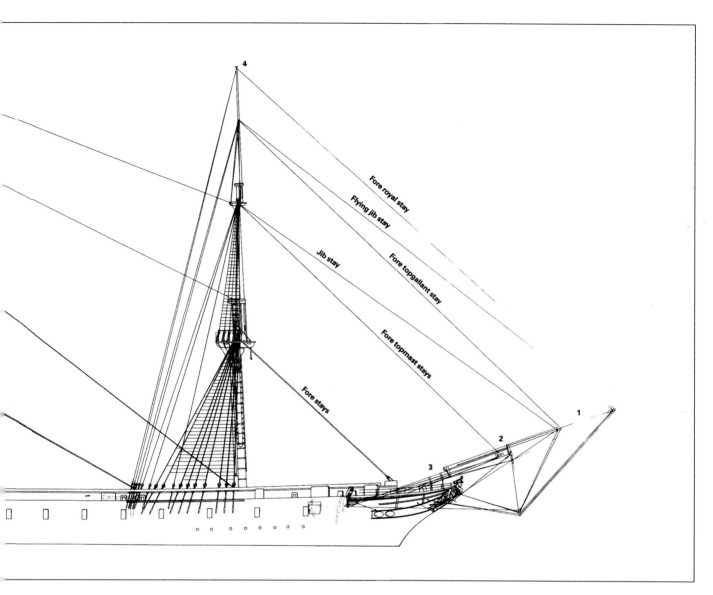

Fore royal stay

Flying jib stay

Jib stay

Fore topgallant stay

Fore topmast stays

Fore stays

the masts presents a seemingly tangled web to the layman. Every rope has a name unique to sailing ships; halyard, guy, stay, sheets, shroud, backstay, tie, pendants, buntline, topping lift, the list goes on. . . .

All this complexity starts to make sense when you separate the 'standing rigging' (which are those ropes and wires that do nothing more than support the masts, yards and bowsprit, with shrouds and backstays) from the 'running rigging'.

The shrouds are the numerous lateral ropes that come down on to the bulwarks and have the ratlines crossing them which allowed the seamen to 'go aloft'. Stays are the ropes that support the masts in the fore and aft line.

	Height above deck	Yards
Foremast	169ft (51.5m)	Fore course
		Fore topsail
		Fore topgallant
		Fore royal
Mainmast	175ft (53.3m)	Main course
		Main topsail
		Main topgallant
		Main royal
Mizzenmast	138ft (42m)	Mizzen top
		Mizzen topgallant
		Mizzen royal
		Spanker
Bowsprit		Stay sail
		Jib
		Flying jib

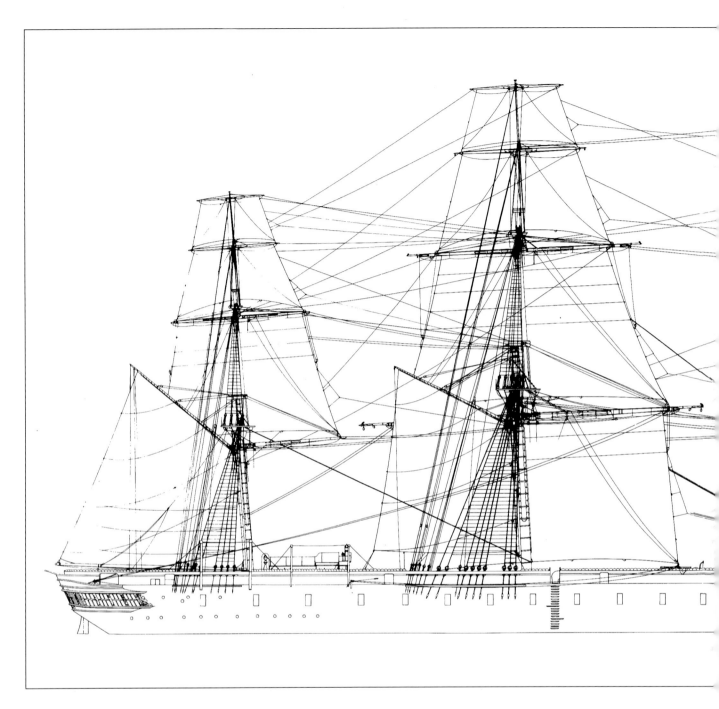

The other ropes are 'running rigging' and as
their name indicates, they are the ropes that
move to raise and lower the sails and yards,
trim the yards into the wind and haul flags and
signals aloft. The yard arms, to which the sails
are attached, could be hauled up and down
their respective masts on what were called lifts.
The yard arms were connected to the mast by
a truss or parrel, which allowed the yard to pivot
towards the wind by using the braces attached
to the end of the yard. The main yard weighed

15 tons alone. Along each yard were stirrups
that supported foot ropes for the seamen when
making sail. The fore and aft sails were hoisted
using halyards and trimmed to the wind by using
sheets (which is yet another name for a rope).

How did *Warrior* perform under sail? At the
time of her construction, a ship's capability was
judged by how well she sailed under canvas;
fighting efficiency and steaming qualities
were incidental. But *Warrior* had broken this
entrenched and seemingly immovable barrier

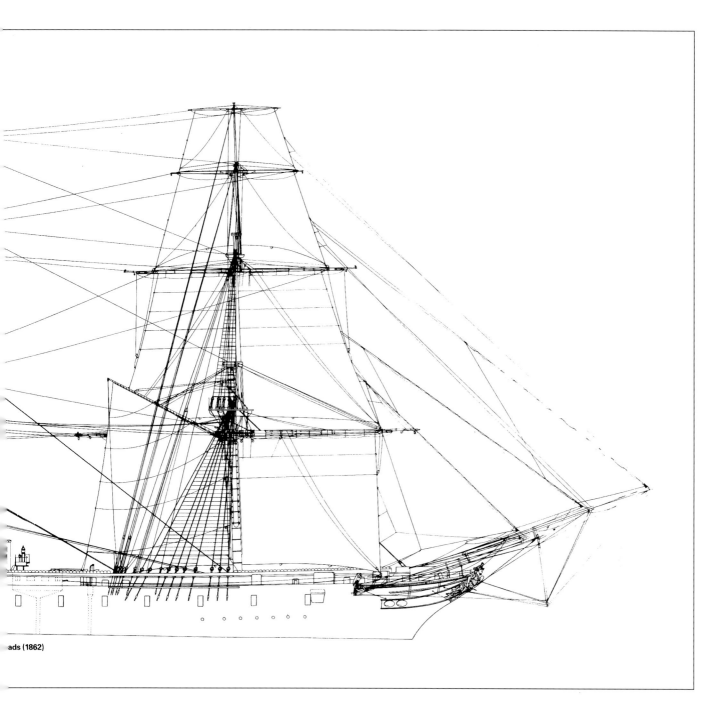

ads (1862)

as steam power was her forte. Many captains, squadron commanders and admirals, however, still insisted that sail came first and it has to be admitted that *Warrior* did not sail well in light winds in comparison to wooden sailing ships. However, the upside was that when wind was strong and all sail set, she was very competent. Having fine lines and being built of iron gave her masts and spars a stiffness that allowed her canvas sails to be held longer into higher winds when other ships would have had to

reef. In March 1864, however, when crossing the Bay of Biscay, she encountered a gale and she broached at right angles to her course. Captain Cochrane recorded that they '. . . held the helm to starboard for ten minutes; ship did not answer helm'. Her small, ineffectual rudder certainly contributed to this situation.

Warrior carried six tapering square sails, three on the foremast and three on the mainmast, a 'fore and aft' sail on the mizzenmast known as a 'spanker' and above

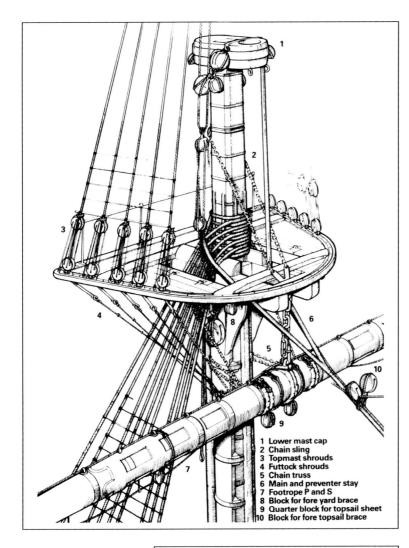

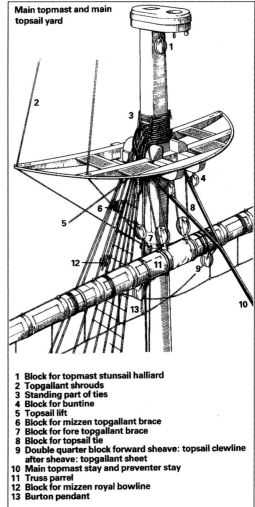

1 Lower mast cap
2 Chain sling
3 Topmast shrouds
4 Futtock shrouds
5 Chain truss
6 Main and preventer stay
7 Footrope P and S
8 Block for fore yard brace
9 Quarter block for topsail sheet
10 Block for fore topsail brace

Main topmast and main topsail yard

1 Block for topmast stunsail halliard
2 Topgallant shrouds
3 Standing part of ties
4 Block for buntine
5 Topsail lift
6 Block for mizzen topgallant brace
7 Block for fore topgallant brace
8 Block for topsail tie
9 Double quarter block forward sheave: topsail clewline after sheave: topgallant sheet
10 Main topmast stay and preventer stay
11 Truss parrel
12 Block for mizzen royal bowline
13 Burton pendant

ABOVE Maintop and yard. (Gary Cook)

ABOVE RIGHT Main topmast and main topsail yard. (Gary Cook)

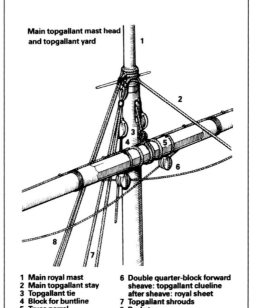

Main topgallant mast head and topgallant yard

1 Main royal mast
2 Main topgallant stay
3 Topgallant tie
4 Block for buntline
5 Truss parrel
6 Double quarter-block forward sheave: topgallant clueline after sheave: royal sheet
7 Topgallant shrouds
8 Backstay

RIGHT Main topgallant masthead. (Gary Cook)

were two square sails. Right forward three jib sails led from the foremast to the bowsprit and jib boom. Between the masts, stay sails could be set. This configuration was known as 'plain sail'. 'Royals', which were smaller sails, could be hoisted at the very top of all three masts.

Under plain sail if the wind increased the first sails to be taken in would be the royals on each mast and the outermost flying jib. If the wind increased further, the topsails would be brought round to 'spill' the wind, reefed to the first line of reef points and then reset to the wind. Further increases in wind would mean further reduction in sail area – of the topsails and possibly the spanker. If the wind continued to increase, a progressive reefing of sails would continue with the topsails being fully reefed and then the mainsails being taken in a reef at a time.

Warrior sailed best when the wind was abaft the beam, to the side of the ship slightly

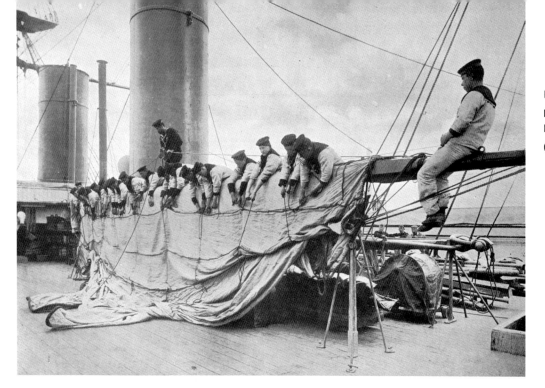

to the aft end. When sailing to windward, the ship would need to tack and this was a careful manoeuvre involving easing the helm slowly down to ensure the sails were kept full by letting go of the headsail sheets that turned yards as the ship manoeuvred. The large spanker sail would be brought amidships to aid the steering. As the ship turned away from the wind, the sails were 'taken aback' – no longer filled with wind from the stern they were now having wind on to their face, which would bring the ship to a slow speed but with the yards

set would aid in swinging her round. To wear ship – turning the stern to the wind – was a longer process and could take anything from 15 minutes up to an hour.

Warrior carried three sets of sails, known as 'suits'. Two would have been regularly used and the third as a reduced set less frequently used. Sails were normally furled on to their respective yard arm. Others were stowed in the sail room where the sailmaker would carry out repairs when necessary. Large sail lockers are located amidships on the lower deck.

BELOW *Warrior*'s sail plan. (WPT)

1 Flying jib
2 Jib
3 Staysail
4 Fore course
5 Fore topsail
6 Fore topgallant
7 Fore royal
8 Main course
9 Main topsail

10 Main topgallant
11 Main royal
12 Spanker
13 Mizzen topsail
14 Mizzen topgallant
15 Mizzen royal

RIGHT Engine room telegraph. *(Author)*

RIGHT The engine builder John Penn. Between the original name is that of Jim Wilson who rebuilt the engines during *Warrior*'s restoration. *(Author)*

RIGHT John Penn, engine builder. *(Public domain)*

Steam

Warrior's engine is known as a Penn 5,400 indicated horse power (IHP), twin cylinder, double acting, single expansion, horizontal trunk steam engine. Quite a mouthful!

John Penn senior started as a manufacturer of agricultural appliances and corn-grinding machinery in the late 18th century. The business was situated at the junction of the Blackheath and Lewisham Roads (adjacent to modern-day Deptford Bridge). His son, John Penn junior, began work with his father and in due course took over the running of the business in the 1840s, when it became John Penn and Son. The younger John Penn was an innovative engineer and began building marine engines, his early models being beam engines of 40hp fitted to paddle-steamers. His later models had twice the power output of existing engines, yet took up a smaller space when compared with those manufactured by Bolton and Watt or Maudsley, companies which had at that time monopolised the marine engine supply business.

Trunk engines were vertical when first produced, but Penn quickly realised that mounting the engine sideways reduced the height dramatically and at the same time retained the long stroke of the piston. This also positioned the engine lower in the hull below the waterline, so that when installed in warships protection was afforded from enemy fire. It also lowered the ship's centre of gravity, which in turn improved stability and seakeeping.

As a result Penn became pre-eminent in the design and manufacture of marine steam engines. He had formed a close working relationship with Joseph Whitworth, who was famous for introducing engineering standards and accurate measuring instruments to enable mass production of quality engines.

Penn's introduction of the trunk engine driving a screw propeller in about 1846 took the business to new heights. Over 230 trunk engines were manufactured. Even the SS *Great Britain* was re-engined with Penn engines after ten years of service. They gained an enviable reputation for quality and reliability.

From Penn's inventive mind came his use of lignum vitae, a dense Caribbean hardwood

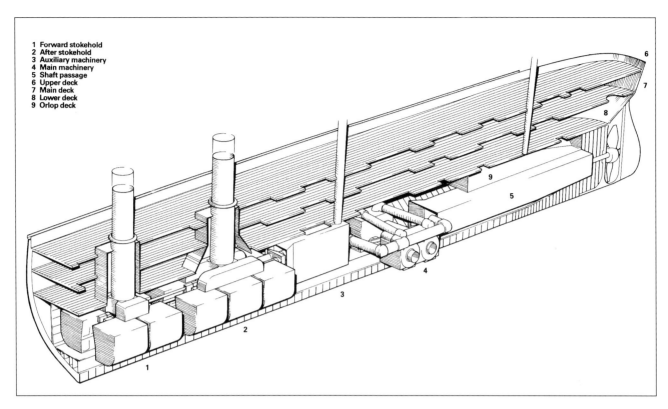

1 Forward stokehold
2 After stokehold
3 Auxiliary machinery
4 Main machinery
5 Shaft passage
6 Upper deck
7 Main deck
8 Lower deck
9 Orlop deck

which he used for propeller shaft bearings and in the stern gland (the watertight seal around the propeller shaft where it emerges from the ship's hull). A unique hardwood that sinks in water, it was used in ships up to the 1970s, only being replaced when synthetic seals and self-aligning ball and roller steel bearings were introduced.

Technical details

A trunk engine has the connecting rod within a large-diameter hollow cylinder known as the 'trunk', the advantage being that the trunk carries little load. The interior of the trunk is open to the air, and accommodates the side-to-side motion of the connecting rod, which links the piston head to the crankshaft. The crankshaft is connected to the propeller shaft.

The working portion of the cylinder is annular (or ring-shaped), with the trunk passing through the centre of the cylinder. Trunk engines relied on high-volume steam at low pressure. *Warrior's* engines had a maximum working pressure of 22psi (1.5 bar). The engine is a twin cylinder with each cylinder 112in (2.8m) in diameter with a stroke of 48in (1.2m). Steam was let into the cylinder on either side of the stroke, making it

a double-acting engine. In simple terms, steam drove the cylinder backwards and forwards. It is a single-expansion engine where the steam is used only once.

The steam was known as saturated steam or 'wet' steam. From the boilers it passed into a steam condensate separator where the water,

ABOVE Layout of boilers and engines. *(Gary Cook)*

LEFT Steam condenser. *(Author)*

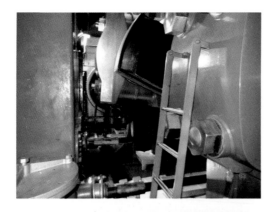

once removed, simply fell to the bottom of a large chamber before the steam passed into the valve chests.

On top of the engine is a double-ported slide valve, which is controlled by the steam valve control gear. This manages the amount of steam that enters each cylinder and can

BELOW **A view at the rear of the engine showing the trunk cylinders.** *(Author)*

bring the engines to a stop or into reverse operation. When the valve opens, it allows steam into the cylinder but the critical aspect of the valve is when it closes; known as the 'cut off', as it closes the port, the pressurised steam is trapped inside and expands, moving the cylinder. It very quickly loses its pressure as the cylinder moves.

At maximum steam pressure with all ten boilers in use the engine produced 5,269 indicated horsepower (ihp) (3,929kW). However, most of this effort was expended in just turning the engine and only 1,250 nominal horsepower (nhp) (932kW) reached the shaft.

After it had done its job, the steam was exhausted into the condenser, which is a large tank on the starboard side of the engine room with a walkway on top of it. A partial vacuum was created within it using an air pump. Sea water was pumped on to a cooling plate within the condenser, causing the steam to condense and collect in the tank. This tank was manually monitored and when necessary the water was pumped back into the boilers or overboard.

The crankshaft and the engine were dependent on good lubrication and all the moving parts have brass lubrication boxes permanently attached. These were filled with tallow, an animal fat, which was melted by the heat of the engines, and therefore lubricated the bearings. The smell can be left to the imagination. At the aft end of the crankshaft at the rear bulkhead of the engine room is a large gear wheel. This is the shaft turning gear.

To enable the propeller to be lifted, it was necessary to ensure the propeller blades were in the correct orientation. Marked on the face of the gear wheel is a white line that when vertical indicates that the propeller blades were also vertical. It was impossible to bring the engines to a precise stop, so a capstan-driven worm drive was lowered on to the large gear wheel and manually turned until the white line was vertical.

In service, the engineering staff measured and recorded the performance of the propulsion machinery with meticulous regulation and precision. These covered propeller rpm, steam pressure, bearing temperatures, coal consumption, boiler water salinity and the maintenance that had been carried out.

When new, *Warrior* conducted a full power

speed trial and achieved 14.3kts. This was an unprecedented speed for a ship of her size and many years passed before another vessel bettered her. With great credit to Penn and his engines, *Warrior* could still achieve her maximum speed to the end of her seagoing life.

In the first two commissions *Warrior* used steam propulsion 75% of the time with either four or six boilers lit; this gave a speed of between 6 and 8kts with a shaft speed between 20 and 30rpm. This was called 'Cruising Speed'. Coal consumption was between 2 and 3 tons per hour. Even when under sail two boilers always remained lit. When she was first built, it was the practice to achieve maximum steam pressure of 22psi (1.5bar) and then control the power with the throttle valve, but this quickly changed to controlling the engine power by varying the boiler pressure so at cruising speed boiler pressure was between 12 and 17psi (0.8 and 1.1bar).

Boilers

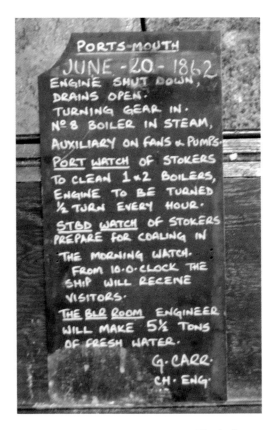

LEFT Engine stateboard showing the activity for the day. *(Author)*

Warrior has ten boilers in two boiler rooms which were known as 'stoke holds'. Designed and built by John Penn, they provided steam to his engine. These were smoke tube boilers as opposed to water tube boilers.

There are six boilers in the aft stoke hold and four in the forward stoke hold. The boilers are of simple riveted rectangular construction from wrought iron, each containing 440 brass smoke tubes. Each boiler was 14ft wide × 12ft long × 12ft high and held 17 tons of salt water.

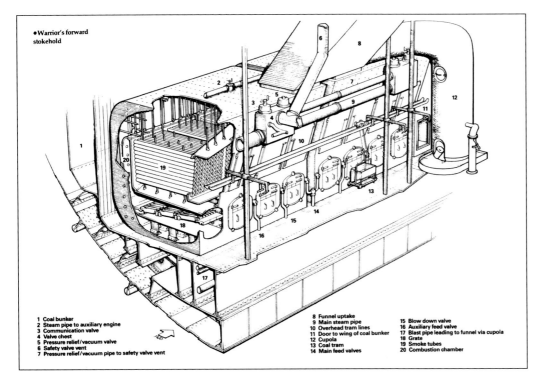

• Warrior's forward stokehold

1 Coal bunker
2 Steam pipe to auxiliary engine
3 Communication valve
4 Valve chest
5 Pressure relief/vacuum valve
6 Safety valve vent
7 Pressure relief/vacuum pipe to safety valve vent
8 Funnel uptake
9 Main steam pipe
10 Overhead tram lines
11 Door to wing of coal bunker
12 Cupola
13 Coal tram
14 Main feed valves
15 Blow down valve
16 Auxiliary feed valve
17 Blast pipe leading to funnel via cupola
18 Grate
19 Smoke tubes
20 Combustion chamber

LEFT *Warrior's* forward stoke hold.

(Gary Cook)

Beneath each boiler are four furnaces and under these were the ash pans.

The performance figures are:

Boiler steam pressure max	22psi (1.5 bar)
Maximum fuel consumption	8.8 tons of coal per hour
Efficiency	67%

On reflection, it is surprising that higher boiler pressures were not provided as by the 1850s railway engines with cylindrical boilers were running at 100psi (6.9 bar). But marine steam engines, epitomised with Penn's trunk engines, were reliable and required minimal maintenance. Propeller design at this time was based on a large propeller producing thrust at relatively slow rotation.

The coal was brought to the firing aisle by the coal trimmers and the stokers fed the furnaces. The temperatures that these men worked

TOP The aft stoke hold. *(Author)*

LEFT Stoke hold with door open on a fire grate. The rectangular barrow delivered coal to the furnace face. *(Author)*

BELOW LEFT View into the fire grate. The tools were used for various tasks, from levelling the coal bed to keeping the ash grillage clear. Below the grate is the ash pan. *(Author)*

BELOW A coal barrow brought coal from the bunkers, which were behind the boilers. *(Author)*

in ranged from 84°F (29°C) to 129°F (54°C) maximum. One observer who braved the stoke hold heat when all boilers were lit described the stokers' actions as almost balletic when firing the furnaces. As the furnace door was opened, one stoker held the blade of his shovel in the face of his colleague to deflect the heat, who in turn shovelled the coal into the fire. It was essential that the coal was evenly spread, creating a bed. A two-pronged rake was used to achieve this. The engine log shows that the use of all ten boilers together was rare and only fired after refits or for speed trials with other steam ships.

Boilers lit	Speed kts	Speed mph	Speed kph
10	14.3kts	16.5mph	27.5kph
6	11kts	12.6mph	18.6kph
4	7kts	8mph	12.9kph
2	0	0	0

It is perhaps surprising that the boilers used sea water, especially as the brass smoke tubes were continuously eroding due to electrolysis, but sea water was readily available. The salinity of the boiler water was continuously monitored and condensed water fed back into the boilers to reduce it, but it was always a saline liquid and the replacement of eroded smoke tubes became a regular and accepted procedure. The boilers themselves were rebuilt at the first refit.

Coal

Coal is a fuel that is disappearing from our modern environmentally conscious times, but it heated us and cooked our food for hundreds of years. It was the fuel of the Industrial Revolution that drove us into the modern world and made Britain a global powerhouse.

Warrior carried 850 tons of coal in six large and two small bunkers outboard of her boilers. This was another innovative idea – using the bunkers as protection of the vital machinery spaces. The coal provided sufficient fuel to drive the ship 1,400 miles at 12kts or 3,500 miles at 6kts.

Working inside the bunkers were men known as coal trimmers who, and by their very name, had to ensure that coal was kept evenly distributed in each bunker to keep the ship evenly 'trimmed' and not take up a list if too much coal

was taken from only one side. The coal was kept permanently wet as there was full awareness of the hazards of coal dust explosion and lethal methane gas, known as fire damp. There were no lights in the bunkers and the trimmers worked with only a small miner's lamp. They shovelled the coal into tubs on a tramway that was pushed to the end of the firing aisle where it was transferred into another tub that was suspended from an overhead rail. This ran along the face of the furnaces where it was dumped on the deck to allow the stokers to feed the fires.

The cost of coal was dictated by how far it had to be shipped. At Plymouth, for instance, it was 60p a ton, but at Madeira it was £1.40 per ton. There was great effort to ensure that the coal came from a very specific source, namely South Wales, and preferably the Nixon Duffryn Colliery in the Aberdare Valley. This was the deepest coal mine in Wales and it produced anthracite, a harder version of coal that burned hotter and more efficiently leaving only a fine ash that fell through the fire grates into the ash pans. The ash that had collected in the ash pans had to be regularly raked out on to the firing aisle floor which was also kept permanently wet. From there the ash was shovelled into the ash bucket that was suspended on the overhead rails. The bucket was then traversed to the forward end of the stoke hold and hoisted to the upper deck. To the port side of the forward funnel is another overhead rail and the bucket moved to a hatch set into the bulwark from where its contents were discharged into the sea.

Getting the coal into the ship, known as

BELOW LEFT Coal bunker stateboard showing the layout of the bunkers and their capacity. *(Author)*

BELOW Coal stateboard recording what was used per day. *(Author)*

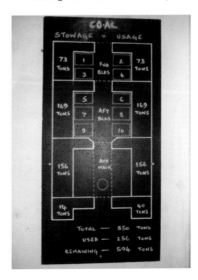

ABOVE View into the boiler casing showing both funnel uptakes and ventilation ducting bringing combustion air into the stoke holds. *(Author)*

been pulled inboard and the mess furniture stowed). Along the gun deck are coaling scuttles through which the coal was shovelled down canvas chutes and into the bunkers. On reaching its destination, the coal was evenly distributed by the coal trimmers.

At each coaling *Warrior* embarked about 400 tons. At best the crew could load 24 tons in a day, so coaling ship was certainly not a quick evolution. Likewise cleaning the ship and crew post-coaling was another unenviable task, bearing in mind that the decks had been holystoned. Some even claim that they were almost white. They certainly would not have been pristine after coaling, so the cleaning started again. The captain, however, had the authority to issue an additional rum tot on completion, which must have been a small incentive at least to get the job done.

Auxiliary machinery

*W*arrior was unique in many ways and one of the most significant for a ship of this period was the 16 watertight compartments that her hull was divided into. These had evolved over the active life of the ship with modifications and additions being made at each refit. Sliding watertight doors were fitted to a number of bulkheads on the lower deck

'coaling', was not the most popular of duties for the crew. Only the captain, commander and master were excused from coaling. Every other member of the crew had to take turns in the laborious and dirty job of filling the bunkers with coal. The exception may have been the musicians who might instead have provided accompaniment to this tedious task.

Donning their coaling rig – usually their oldest garments – half the crew would go into the collier which had moored alongside and bag up the coal into 2cwt (100kg)·sacks. These were slung up and into the ship through the gunports and dumped on the gun deck (the guns having

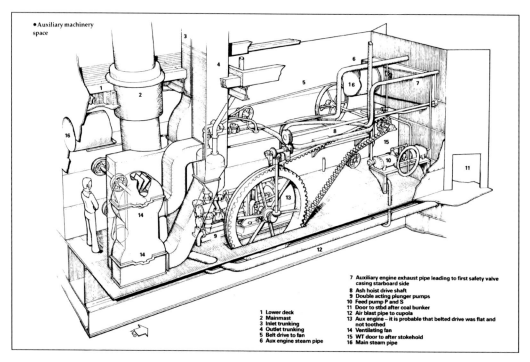

RIGHT Auxiliary machinery space.
(Gary Cook)

• Auxiliary machinery space

1 Lower deck
2 Mainmast
3 Inlet trunking
4 Outlet trunking
5 Belt drive to fan
6 Aux engine steam pipe
7 Auxiliary engine exhaust pipe leading to first safety valve casing starboard side
8 Ash hoist drive shaft
9 Double acting plunger pumps
10 Feed pump P and S
11 Door to stbd after coal bunker
12 Air blast pipe to cupola
13 Aux engine – it is probable that belted drive was flat and not toothed
14 Ventilating fan
15 WT door to after stokehold
16 Main steam pipe

that could be operated remotely from the upper deck. This subdivision of the ship's hull necessitated a considerable and extensive pumping, ventilation and ballasting system.

In naval architecture terminology, 850 tons of coal is known as a 'variable load' and as it is used the ship becomes lighter, affecting the stability and sailing qualities – and potentially the safety – of the ship. Therefore intentional flooding of the double-bottom bilge spaces compensates for this loss, and keeps the ship stable. This is known as ballasting. Likewise, when taking coal on board the water ballast would need to be pumped back out.

At the aft end of the aft stoke hold in the passage leading to the engine room was the 30hp auxiliary steam engine that provided power to the 'double-acting plunger pumps'. These were used to pump sea water into and out of the bilges. It also provided water to the ship's fire-main. Its other tasks included driving the ventilation fans and the ash hoist. In the restoration programme, the auxiliary engine was not replicated.

The piping systems were complex and extensive. They were connected to ship side valves known as 'sea cocks' which, when opened, let sea water into the ship's double-bottom and sluice valves that allowed water to flow between watertight compartments. These were operated by rods attached to turning plates that are situated throughout the ship and in particular on the lower deck. There are round brass plates set into the deck, each engraved with its function.

Despite the power of the auxiliary engine, there was the ability to operate all the piping systems manually using the Downton pumps. This was a positive displacement pump patented in 1825 by Jonathan Downton. Nine are spaced throughout the lower deck and two more are situated on the main deck. Two of the lower deck pumps were fitted with extended gearing up to the main deck where a pair of long crank handles were fitted that up to 30 men would operate.

The pumps not only discharged water from the bilge and double-bottom but also from any compartment that may have been flooded on the orlop deck. The water pumped out would pass into what was known as the port or

LEFT Deck plate in the gun deck for opening or closing one of the watertight doors on the deck below. *(Author)*

LEFT Deck plate that remotely operated one of the numerous valves in the ship. *(Author)*

LEFT The remote operating gear for one of the Downton pumps. The holes in adjacent pillars are where crank handles would have been inserted. *(Author)*

deck is one of the
downtakes to the
boiler room with
the grill and shutter
that was installed to
improve ventilation
for the pipe-smoking
crew. *(Author)*

starboard common sewer system and would
then be discharged into the sea.

Natural ventilation was provided around the
ship via pipes and ducting and down to the shell
room, provisions, spirit and bread stores. Forced
ventilation to the magazines was via manually
operated air pumps. Engine room ventilation
came from the upper deck cowls that forced air
down by fans driven by the donkey engine.

The cable bitts (bollards) that were such
a dominant feature on the cable deck and
half deck had ventilated slots around the top
periphery and as the bitts were hollow this
provided ventilation down to the lower deck.

RIGHT Water tank in
the cell flat. *(Author)*

RIGHT Water tank on
the gun deck on the
forward bulkhead.
(Author)

On first commissioning the main deck that
accommodated some 650 men had not been
provided with any ventilation. The gunports
remained tightly closed, only being opened for
gunnery drill. As nearly all the crew smoked pipes,
a 'fug' quickly formed. It was claimed that if one
stood at the citadel entrance the far end of the
main deck could not be seen so dense was this
tobacco smog. In an odd way the crew liked it.

However, condensation off the iron structure,
exacerbated by the regular scrubbing of the
decks and inadequate scuppers, produced a
permanently damp atmosphere and serious
health issues became prevalent. To alleviate this,
modifications were made to install two ventilation
outlets in the downtake ducts to the engine
room and the stoke hold with a sliding shutter to
control airflow on to the main deck.

Additional scuppers were also installed to
drain away water. When the ventilators were
installed the crew complained about 'all the
unhealthy fresh air', but it had the desired effect
of dramatically improving the air quality. There is,
however, some anecdotal evidence to suggest
that the shutter was closed more often than
open to maintain the 'fug'! The ventilator also
had an additional function. When fully opened,
it slightly pressurised the main gun deck so that
when the guns were fired the smoke was driven
out through the gunports and not back into the
ship – Victorian ingenuity at its best.

One of the most significant engineering
achievements on *Warrior* is one that is not
apparent to the visitor but of huge significance
to the crew. It was the availability of almost
unlimited quantities of fresh water. There are 100
tanks containing over 100 tons of fresh water on
the ship. The majority of the tanks are positioned
around the forward and aft magazines with
'ready-use tanks', which were filled by hosepipe,
located on the main and lower decks for
drinking water, washing and cooking purposes.
The ship consumed about 5 tons of water a
day. The water was initially supplied from ashore
but the ship had two Grant's steam condensers
located above the forward magazine which
could make about ¾ton of fresh water an hour.
The ship's records show that between 5 and 20
tons a day was produced.

In the forward stoke hold there is a furnace,
known as a cupola. This is a vertical cylinder of

iron lined with fire bricks that exhausted into the forward funnel. The purpose was to heat metal to a molten state to fill the Martin's Liquid Iron Shells for the 68lb cannon (for more information see Chapter 3 – Firepower). It also provided the blacksmith with metal for forging.

With wooden warships, the maintenance of the hull and the fixtures and fittings was a task for the shipwrights, whose tools would have been chisels, adzes and saws. With *Warrior*'s iron hull brought changes and new skills requirements. On the ship is a range of machinery and tools including anvils, forge, metal-cutting lathes that could produce screws and shearing and cutting machines. The ship carried a huge range of spares, lubricants and even cloth from which replacement clothing could be manufactured, especially vital for the stokers and coal trimmers. Even a spare propeller blade was stowed in the laundry. This followed the Royal Navy policy that their warships should be as self-reliant as possible when it came to maintenance and support, given that many were deployed at the far reaches of the British Empire with little access to dockyards. Likewise, when the ship saw action, the crew should have the ability and the means to repair battle damage.

FAR LEFT Spare propeller blade stowed in the laundry. *(Author)*

LEFT The foundry in the engine room. *(Author)*

The shaft

The power that the engine produced was transmitted to the propeller by the propeller shaft. The shaft itself represented engineering of considerable ingenuity for the time. When one looks at *Warrior*'s plans, the engine is almost in the middle of the ship and consequently the shaft needed to be exceptionally long.

Not only did it have to be sufficiently strong to endure the axial loads of pushing the propeller round but also to accommodate the thrust that it produced along with a need to pass out of the hull of the ship with a watertight seal. Just to add further complexity, there was a requirement to disconnect the propeller from the shaft to enable it to be lifted.

The shaft was of solid wrought iron 108ft long × 17in diameter and comprised of five lengths each bolted together. The whole weighed nearly 55 tons, along with its four supporting pedestal bearings known as

RIGHT A part of the original shaft remains where it passes out through the hull. *(Author)*

plummer blocks. The forward one was the thrust block that, as its name suggests, absorbed the load of the propeller driving the ship forward. The shaft tunnel itself is large at 7ft × 7ft with the ship's keelson running its full length. Due to access difficulties, the tunnel is not open to the public and the shaft itself has not been restored. Only one portion of the original remains.

The propeller

As to who by and when the invention of the marine screw propeller was made is somewhat lost in history and cannot be attributed to a single individual – despite many claimants. As early as in 220 BC Archimedes understood the principle of a screw-shaped shaft that was used to raise water for irrigation. Much later the Scottish engineer James Watt promoted the idea for a propeller for marine use in 1770 but refused to allow the use of his steam engines on a ship.

It was only when steam engines were installed in ships, or more probably barges, were the minds of numerous inventors focused as to what might be the most efficient means of propulsion.

In 1835, two inventors – John Ericsson, a Swedish national working in Britain, and Francis Pettit Smith – began working separately on finding the most efficient propeller. Smith was a farmer by trade who had a lifelong fascination with screw propulsion and was first to take out a screw propeller patent in May 1835. Ericsson's work resulted in him filing his patent six weeks later. Both built small boats to prove their designs. Ericsson demonstrated his to an Admiralty committee who rejected it out of hand, seeing it as impractical for oceangoing vessels on an ill-founded assumption that propeller-driven ships could not be steered efficiently. Ericsson's response was to build a second, larger screw-propelled ship, the *Robert F. Stockton* that sailed to the United States in 1839 where he was to gain fame as the designer of the US Navy's first screw-propelled warship, USS *Princeton* in 1843.

Prior to this, the SS *Archimedes* is recognised as the first oceangoing propeller-driven ship. A wooden schooner of 240 tons she was built on the Thames and first went to sea in 1839 fitted with a propeller designed by Francis Pettit Smith. She had a great influence on Isambard Kingdom Brunel when he was designing the SS *Great Britain*, which was eventually built in 1845.

As with all inventions, it is sometimes those who take an existing idea and have the ability to develop it to its full potential who achieve the greater success. Robert Griffiths is attributed with bringing screw propellers to an efficient and practical level and with it an understanding of the mechanics and hydrodynamics of the propeller. As such, in many journals and documents of the period he is attributed with having invented the modern propeller.

Griffiths came from the Vale of Clwyd in North Wales and showed an early inclination for mechanical pursuits – in particular wood turning. Working for an engine builder in Birmingham, he began to file patents for riveting machines, glass grinding and polishing machines, railway spikes and nut and bolt manufacturing machines. He later began his own business in France, but that was brought to an abrupt end under Napoleon III's oppressive reign and he returned to Britain.

Re-establishing his business in Britain, he started experimenting with improving the screw propeller and he filed a number of patents that made propellers more efficient. This was possible owing to his understanding of blade pitch angle and the balance of the propeller that reduced vibration. He also established the distance a propeller needed to be from the hull to the hub and the idea of having separate

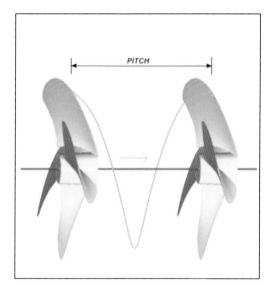

PITCH

blades that could be unbolted from the hub making blade manufacture and repairs simpler. Most significantly, he proved that just two large blades were as efficient as the multi-blade propellers that were being used on other ships.

Thomas Lloyd, the Admiralty's Chief Engineer, had himself studied propeller design and he saw that Griffiths' designs were based on sound principles; they were adopted not only for *Warrior* but also for many subsequent vessels.

'Up screw, down funnel'

In the 1850s when the plans for *Warrior* were being prepared, there were innovative aspects of her design that carried significant technical risk. Griffiths' two-bladed propeller was one of these risks. Not only did it have to drive the ship forward, it also presented a potential problem when it was not turning. The mindset of the Admiralty Lords was to assess the quality of a ship based on how well she performed under sail. From experience gained from other ships, a static propeller caused significant hydrodynamic drag that seriously affected a ship's sailing performance. With this in mind, *Warrior* had built into her design the ability to lift the propeller clear of the water. Once again, Victorian ingenuity solved a potential problem.

Visitors to today's ship can see little of the engineering for raising the propeller. The exception is on the upper deck at the aft end adjacent to the 110lb chase gun. Within the

RIGHT Propeller-lifting well cover plate. *(Author)*

FAR RIGHT A view into the lifting well. The lifting mechanism has not been restored and the well is empty. *(Author)*

RIGHT The lifting well behind the captain's cabin. The steering ropes pass around the well. *(Author)*

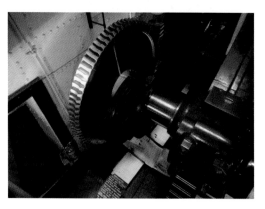

FAR RIGHT Engine room showing propeller alignment gear. The white line indicates that the propeller blades are vertical in its lifting frame. *(Author)*

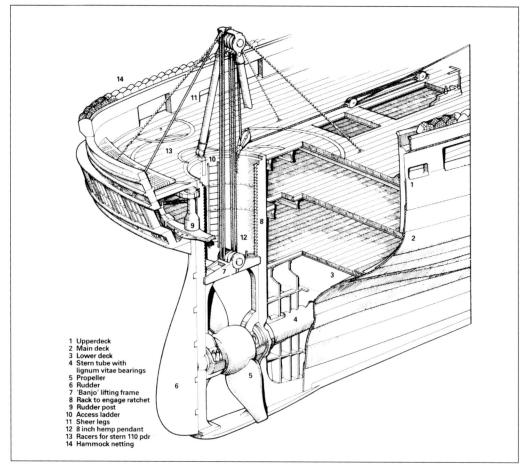

1 Upperdeck
2 Main deck
3 Lower deck
4 Stern tube with lignum vitae bearings
5 Propeller
6 Rudder
7 'Banjo' lifting frame
8 Rack to engage ratchet
9 Rudder post
10 Access ladder
11 Sheer legs
12 8 inch hemp pendant
13 Racers for stern 110 pdr
14 Hammock netting

RIGHT Propeller lifting gear. *(Gary Cook)*

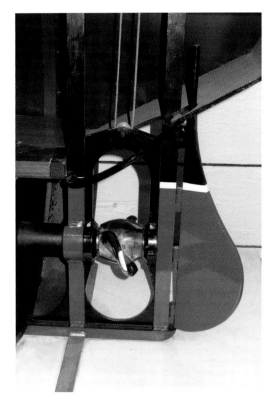

**FAR LEFT Model
showing the banjo
lifting frame.** *(Author)*

**LEFT The propeller
partially lifted into the
well.** *(Author)*

brass races for the gun is a large brass circle that was the lift-out panel above the propeller-raising well. With the panel removed one could look down a wide shaft that was open to the sea and at the bottom would be the propeller mounted in what was known as a 'banjo' frame. The frame itself closed over either side of the propeller on to the emerging propeller shaft. The whole assembly weighed over 35 tons. The top of the frame engaged in ratchet bars that ran up each side of the well that prevented the propeller and frame from falling back if the lifting tackle failed. It was necessary to have the propeller blades vertical to enable it to be lifted and to achieve this shaft turning gear was installed in the engine room that was manually rotated.

At the same time, shear legs would be erected on the upper deck on either side of the open lifting well and hemp rope tackle, known as a 'pendant', was connected to the banjo frame. Here recorded history becomes a little contradictory as to how many men it took to lift the propeller. Some say 400, but one account says 600 men were needed. Whichever is correct, this was clearly a major labour-intensive activity.

With the propeller blades vertical, the propeller hub could be lifted off the propeller shaft as the connection was a simple tongue and groove. When the rudder and frame had lifted to its maximum height, a crewman went down the ladder inside the well and secured the locking plate. It quickly became apparent that this was a task that could only be done in harbour or at anchor in still water. To attempt it in any sea state would be hazardous in the extreme.

How many times did they lift the prop?

In October 1861, while berthed in Portsmouth, the propeller was lifted and engineers from John Penn changed a blade using the spare stowed in the laundry. There is some evidence that the exercise was carried out again just for practice.

It quickly became apparent, however, that with modern ships such as *Warrior* the ability to have steam propulsion readily available at all times had become essential when at sea. There was also a concern that the shaft connection which the propeller dropped into could become quickly fouled by marine growth if left open for any length of time, preventing the propeller being reinstalled. Even when under sail, the boilers were never fully shut down. A minimum

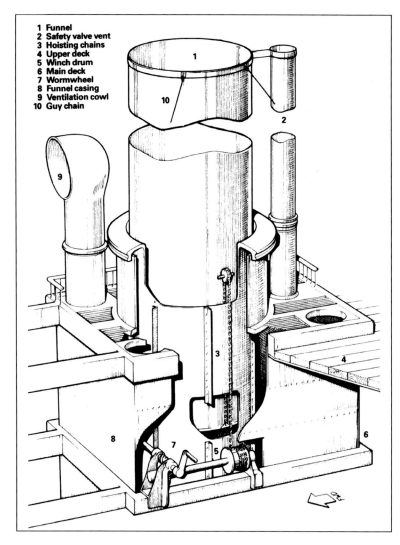

1 Funnel
2 Safety valve vent
3 Hoisting chains
4 Upper deck
5 Winch drum
6 Main deck
7 Wormwheel
8 Funnel casing
9 Ventilation cowl
10 Guy chain

ABOVE Funnel hoisting gear. *(Gary Cook)*

BELOW The armoured box. *(WPT)*

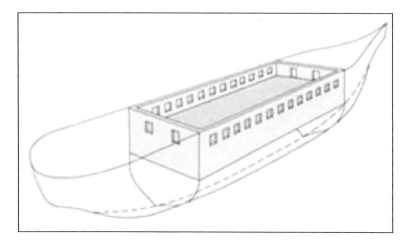

of two would always remain lit. Therefore, with steam being produced – albeit in low volume – it was sufficient to turn the engine and propeller drag became negligible. Propeller lifting as an evolution disappeared.

As with the propeller, the large funnels had been perceived as having a detrimental effect on the sailing qualities of the ship. Again this proved a misconception and, though the funnels could be lowered by a winding mechanism within the funnel casing, after a short period of service it became another evolution that was no longer required.

There is some anecdotal evidence that in *Warrior*'s early days, admirals inspecting ships from ashore or from the Admiralty yacht considered funnels not aesthetically pleasing (their preferences lay firmly in the age of sail), and they ordered the funnels lowered.

Armour

The armour plating is the most significant of the many technological advances that can be found on *Warrior* and is at the very heart of her strength. Over 213ft of the midship length of the ship is armoured; the two transverse bulkheads and from the upper deck to 6ft below the waterline to form a box. It became known as the 'citadel'.

The Industrial Revolution in Britain was accelerating at an unprecedented pace in the production of iron, the manufacture of more powerful cannon and the imminent prospects for the new large-rifled guns. The approach to the armouring of the ship had to take these emerging factors into account. What would it have to stop? Could effective armour be manufactured in quantity and quality? How heavy would the resultant armour be? One of the critical design decisions was not to armour the bows and stern of *Warrior* as it could have adverse effect on both immersion and stability owing to the additional weight.

The ship's structure is wrought iron as merchant ships of the period were. The armour is likewise wrought iron. Cast iron had been used as armour protection in some land fortifications but wrought iron was found to be superior for naval use. The differences between these two materials is the carbon content which makes cast iron very hard but brittle and liable to shatter when struck. Wrought iron remains malleable and is more easily worked into various shapes during the smelting process.

Firing trials had been conducted on gunnery ranges for some years, culminating in an extensive series of tests on the ranges at Shoeburyness on the Essex coast. Different types of guns had been fired against various simulated targets that represented wooden ships, iron ships and armoured ships. The 68lb (31kg) cannon destined to be carried on *Warrior* was fired against wrought iron armour. It was discovered that 4in (100mm) of wrought iron armour would stop a 68lb ball at maximum velocity. From these trials, the decision was made to construct Warrior's box from 4½in (114mm) wrought iron armour plate. The extra ½in (12mm) thickness was for added assurance.

There were two processes for manufacturing iron armour. Basic iron plates were manufactured using the 'puddling' process – simply pouring liquid metal into an open sand mould – but this process could only produce reasonable quality iron up to about ½in (12mm) thick. To achieve the required armour thickness, ½in plates were stacked together along with scrap iron that had been rolled into bars and the whole lot was reheated to a high temperature. Considerable skill was needed in judging the correct temperature that was 80–90% of the melting temperature. The heated stacked plates were taken from the furnace and hammered using very large steam hammers which welded the plates together into one solid, thick plate. The Thames Iron Works had acquired a considerable number of 4-ton Nasmyth Gaskell steam hammers and the men who operated these machines were known as hammersmiths. Subsequently this area of the works was known as 'Hammersmith'.

The second process was rolled iron, which again followed the practice of stacking plates and scrap iron bars together and heating them to a high temperature, but this time, rather than hammering, it passed between iron rollers to become one plate. This produced better and more consistent quality iron but at the time of *Warrior* it was difficult to manufacture. Rolling machinery of exceptional size and power was required to produce it, and it was not until well into the 1860s that rolled iron became commercially available. It was, however, used on later warships.

Steel was being manufactured and was tested as potential armour to resist shot. The steel being made at this time, though, proved too brittle to be effective and had similarities to cast iron in that it shattered when struck. It was not until the late 1870s that metallurgy advances made steel a viable option. The invention of Henry Bessemer's blast furnace process enabled good quality steel to be manufactured in quantities at commercial prices. It was just too late for *Warrior*.

Each armour plate is 15ft (4.5m) long, 3ft (1m) wide and weighs 4 tons. The ends have a tongue and groove connection so each joined plate gave additional support. This proved a hindrance during the ship's life as it was difficult to extract an individual plate during maintenance.

The gunnery trials had found that iron armour plate would split on the rear face when heavily struck by a shell or round shot. This was known as 'spalling'. Wrought iron, when viewed through a microscope, has a grainy consistency similar in appearance to wood grain. On being struck, the grain separates on the reverse face producing a splintered and potentially lethal shard.

The trials had established that wood installed directly behind the armour prevented spalling, cushioned the shock of a hit from damaging the structure of the ship and distributed the force over a larger area which prevented penetration.

Warrior's final armour installation comprises of 4½in (114mm) of wrought iron armour backed with two layers of 9in thick teak secured to the 1in thick iron hull. The gunports are protected by a 27in (309mm) deep frame of 4½in (114mm) wrought iron.

LEFT A cutaway portion of *Warrior*'s armour. The ship's iron hull is to the left, the two baulks of 9in teak in the centre and the 4.5in of armour is to the right painted white. *(Author)*

RIGHT The first
longitudinal frame
assembly and lower
front of armour plating.
(Gary Cook)

FAR RIGHT Bulkhead
armour. The bulkheads
at the citadel ends
have only 9in of teak.
(Author)

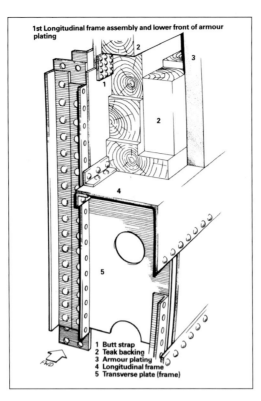

1st Longitudinal frame assembly and lower front of armour plating

1 Butt strap
2 Teak backing
3 Armour plating
4 Longitudinal frame
5 Transverse plate (frame)

FWD

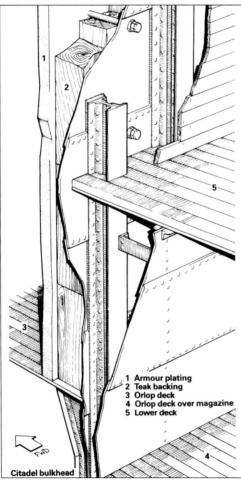

1 Armour plating
2 Teak backing
3 Orlop deck
4 Orlop deck over magazine
5 Lower deck

FWD

Citadel bulkhead

RIGHT Citadel
bulkhead. *(Gary Cook)*

The length of the box or citadel was dictated
by the need to protect the guns, and at first
it was envisaged that 19 guns would form
a broadside, which would necessitate an
armoured box of nearly 300ft. As this was an
innovative build, the serious reservations of
extending the armour to the stern and to the
bows limited the number of guns within the
citadel to 13 each side. Within the armour
protection were seven of the watertight
subdivisions. It was calculated that even if the
unarmoured bows or stern were penetrated and
flooded, they would not endanger the flotation
or stability of the ship.

When completed, *Warrior*'s armour weighed
1,300 tons. This represented $\frac{1}{7}$ of the ship's
total displacement of 9,200 tons.

If you were going to fight a naval battle, *Warrior*
was the ship to be on. She was bullet-proof.

The magazine

Warrior has two powder magazines, one
forward and one aft, positioned low in
the ship well below the waterline. In their design
and construction they had changed very little
from the magazines found on the wooden
warships of the day. The gunpowder was the

most dangerous substance on the ship and with illumination coming only from oil lamps or candles, extreme caution was required when handling and distributing powder around the ship. The magazines are constructed from wood and all fittings are either wood, brass or copper. No iron was permitted as this could cause a spark. The only light came from separate light rooms adjacent to the magazine where bull's eye glass port lights shone candlelight into the magazine.

There was a hand-driven mechanical ventilation system to provide air circulation. Around and above each magazine were 40 of the ship's fresh water tanks, which could flood the magazine in the event of fire. The floor was covered in lead sheeting and the magazine crew wore felt slippers when working in the cramped and dimly lit space.

As described in Chapter 3, different sizes of charge were available and these were made up as requested by the gunnery officer who indicated his requirement by a simple telegraph board. This was operated by rope moving an arrow on to a coloured square, the respective colours indicated the charge size required.

The charges were loaded into leather cases and lifted into the handing room that was

above each magazine and then again up to the gun deck. The empty charge cases were dropped back into the magazine through a canvas tube.

The tubes terminated under the circular white hatches on the gun deck that can be seen adjacent to the armoured bulkheads at either end of the gun deck. Each powder man would take

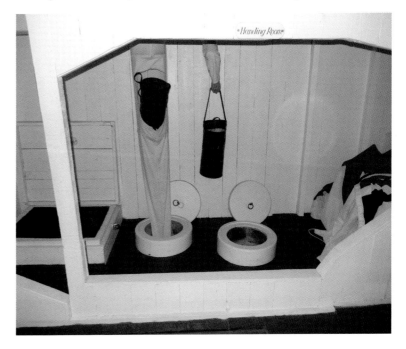

four powder cases at a time and deliver them to the guns. They were expected to deliver no fewer than 60 charges a minute when in action.

Steering

Despite *Warrior* being a technological leap forward in warship construction, she had steering that came directly from the wooden warships of previous generations. As a consequence of this, she had a reputation for poor steering.

This was a failure by Isaac Watts to recognise that on a ship that was 400ft long, a much larger rudder was required. What *Warrior* needed was a balanced rudder that Isambard Kingdom Brunel had developed and fitted to his SS *Great Britain* in 1843, a full 17 years before *Warrior*.

A balanced rudder has the axis of rotation behind its front edge. When the rudder is turned, the pressure caused by the ship's passage through the water acts upon the forward part to exert a force which increases the angle of deflection, so counteracting the pressure acting on the after part, which in turn acts to reduce the angle of deflection. This allows the rudder to be moved with less effort than is necessary with an unbalanced rudder.

Captain Arthur Cochrane recognised this shortcoming on *Warrior* very early in his command and made representation to the Admiralty to have a larger, better rudder fitted. He received only a prevaricating reply and she retained the same rudder throughout her life.

Warrior's rudder post rises up through the stern of the ship, terminating in the compartment behind the captain's cabin (which also contained the propeller lifting well). In this compartment and in the lower deck store, steering arms, known as 'yokes', came off the steering post port and starboard around the propeller well, to which the steering ropes were attached. The ropes themselves were manufactured from boiled leather thongs that did not stretch.

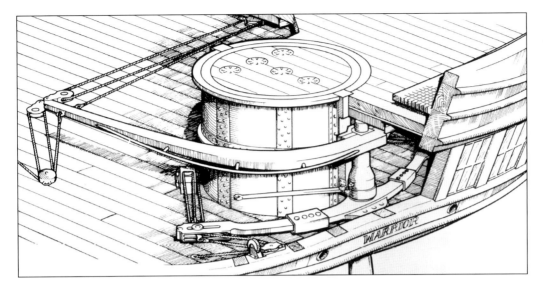

The ropes that can be seen on the half deck emerging from the captain's cabin go to wheels on the upper deck and these were known as the 'cruising wheels'. The ropes from the yokes on the lower deck that can be seen passing through the wardroom run to the wheels that are in the armoured box. These would have been manned if the ship went into action and are consequently known as the 'battle wheels'.

A third set of wheels are located on the lower deck adjacent to the issuing room, but these have no ropes connected to them as they are the emergency wheels and would be rigged as circumstances dictated.

ABOVE This view of *Warrior's* rudder shows just how small it is in relation to the vessel's size. (Author)

On the forward face of the cruising and battle wheels is a rudder angle indicator. The four wheels that can be seen at each of these locations are in reality two pairs with the steering ropes wrapped around the central hub, from which a complex array of ropes, block and tackle moved the rudder. Both sets of wheels turned in the same direction.

The senior quartermaster would control the wheels and keep the ship on the course required. He would stand on the windward side of the front wheel so he could clearly see the compass. As the wind could come from either port or starboard, there are two compasses to enable him to see the ship's heading from either side of the wheel. In fine weather the quartermaster and three men would be sufficient to steer the ship. In worsening weather, up to eight men might be needed. In extreme weather men would be put into the steering spaces and

BELOW The 5.6-ton Bower anchor either side of the bows. (Author)

haul directly on the relieving tackles secured to the tiller arms. Experienced helmsmen would always be on the forward wheels, while trainees would take the after wheels.

Despite Cochrane's pleas for a larger rudder, changes were instead made to the steering gear itself. Very shortly after entering service, the yokes were replaced for 2.5-ton 20ft tillers to improve leverage. They achieved little, however, and considerable physical effort was still required.

In early 1863 Cochrane obtained approval to have 'Renton's Hydraulic Steering' fitted. This comprised of a hydraulic tiller with an actuating cylinder mounted on top of the rudder head with a steam pump, but it was still actuated by ropes from the wheels. Regrettably, by the end of the year Cochrane had to admit that the Renton modifications had not been successful and had contributed little to improving the steering. In February 1864, when she was docked at Plymouth, it was removed and the ship resorted once again to the tiller arms. *Warrior* lived with problematical steering throughout her active life.

Anchors and cable

*W*arrior has nine anchors. Two of her bow anchors were the largest and heaviest that had ever been produced in maritime history. Despite their size and weight, when first built the anchors were deployed and recovered using just manpower.

The anchor that had been used by the navy for many years was known as the Admiralty Pattern, but with the advent of larger warships the Admiralty recognised that other options should be considered as merchant ships were using a number of different designs. A committee was formed in 1853 and various trials were conducted. The results showed that commercial examples were superior in all respects to the traditional Admiralty Pattern anchors. Despite these conclusions, *Warrior's* bow anchors are Admiralty Pattern, comprising:

- 2 × wooden stock anchors of 5.6 tons each. Though both anchors were identical they were known as 'best bower' (starboard) and 'small bower' (port). These were the principal anchors and either one or both could be deployed.
- 2 × wooden stock sheet anchors of 5.6

tons were stowed aft of the bower anchors on crutches that allowed the anchors to be thrown clear of the ship when deployed. They were used as additional holding anchors to the bower anchors.

- 1 × iron 1.4-ton stream anchor used for holding the ship temporarily. It was stowed against the mainmast on the upper deck. It did not always have its stock that was fitted prior to its use.
- 2 × iron kedge anchors of 0.4 and 0.8 tons. These could be deployed by the ship's boats for warping the ship. They were stowed either side of the mainmast.
- 2 × Commander Rogers' patent 3-ton kedge anchors were stowed on either stern quarter to hold the stern, especially when the ship was bombarding.

Cable

In our modern world cable is what is used for transmitting electrical power, but in the navy 'cable' is the word for anchor chain. The ship had chain cable and hemp cable. Cable is measured in fathom lengths (1 fathom = 6ft/2m) and chain cable measured by the diameter of the link, whereas hemp cable is measured by its circumference.

Warrior had a total of 600 fathoms of studded link chain cable 2⅜in in diameter manufactured by Hawks, Crawshay and Sons of Gateshead. Each fathom length was joined together by a joining shackle. The fathom lengths were numbered by having thin wire wrapped around the chain, therefore the number of wire turns indicated how many fathoms had been let go. Some 250 fathoms of chain cable were used for each bower anchor, 100 fathoms of chain cable was used for one of the sheet anchors, whereas the other sheet anchor had 100 fathoms of hemp cable. Each of the stern anchors were provided with 100 fathoms of iron cable.

The chain cable was stowed in four cable lockers located around the mainmast on the orlop deck. The cables were brought up through the lower deck and on to the gun deck via the large black gratings amidships known as naval pipes.

Deploying or raising an anchor was a convoluted, time-consuming and physically

demanding evolution. If a bow anchor was to be deployed, the forward capstan on the gun deck was brought to readiness. The adjacent pillars were knocked out of their shoes and folded up into the deck head. The galley bulkhead would have been folded inward to provide a clear route. The capstan bars inserted into the capstan head were manned by up to 88 men for operation.

Coming off the capstan was a 1⅝in stud link chain that had oblong links that engaged with the sprockets on the capstan. This was known as the messenger cable and was led forward on to the cable deck, passing over the black hooked guides and vertical rollers located along the route and in the doorways. From thence it passed across the cable deck and back along the gun deck on to the capstan, forming an endless loop.

The messenger cable was then 'nipped'

ABOVE Stern anchor – a Rogers' patent 3-ton kedge anchor. *(Author)*

BELOW The top of the cable locker on the gun deck. Cable could be deployed either forward or aft. *(Author)*

to the anchor cable by either iron or hemp nippers. Put simply, this meant that one chain link was tied to another. This practice was traditionally carried out by younger members of the crew as the messenger cable moved all the time and dexterity was needed first to connect the nip and then to remove it before the cable went either overboard or down into the naval pipe. Hence the young crewmen who carried out this task became known as 'nippers'.

Deploying an anchor was easier as the stowed weight of the anchor assisted in its deployment. The hawser plugs, where the anchor cable passed out of the ship and prevented water entering when under way, would be removed. The anchor cable would be hauled forward on to the cable deck and shackled to the anchor to be used, which was held in place by a cable compressor and slip. When fully connected, the compressor was eased off, the slip knocked off and the anchor dropped, drawing the cable behind it. This process was reversed when a stern anchor was being deployed or recovered. The anchor cable passed through the captain's cabin.

Warrior has three capstans, though the two aft capstans can be operated as one. The forward capstan was originally manual but was changed to the present Brown's patent steam-driven capstan during her 1872–75 refit. Aft of the galley is a platform known as the bandstand

LEFT The bandstand
where the musicians
played as the capstans
were turned. (Author)

where the ship's musicians would have played
when the anchors were being worked.

Boats

When *Warrior* first entered service, she was
equipped with nine boats. Again this was
the issue that an 80-gun two-deck wooden
warship would have received.

These were:

2 × 42ft pinnace launches
1 × 32ft pinnace
1 × 32ft galley
1 × 30ft gig

2 × 30ft cutters
1 × 18ft jolly boat
1 × 14ft dinghy

However, in April 1862 Captain Cochrane
asked for changes. He wanted boats with more
oars for greater pulling power. His request was
granted and the outfit of boats is listed in the
table overleaf.

Warships, even in the modern navy, do
not designate their boats as 'lifeboats' but as
ships' boats for general-purpose use. They
transported stores, water and people and have
a military role of delivering landing parties and
ceremonial delivery of the captain and senior

BELOW Ship's cutters
(HMS *Black Prince*).
(Walker Image Archive)

ABOVE The jolly boat at the stern. *(Author)*

LEFT Starboard gig slung from iron davits. *(Author)*

BELOW Ship's launch rigged for carrying an anchor. *(Public domain)*

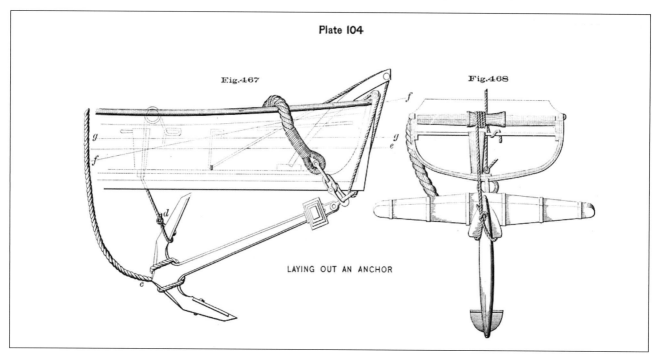

Plate 104

Fig. 467

Fig. 468

LAYING OUT AN ANCHOR

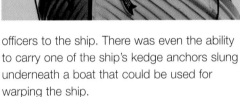

officers to the ship. There was even the ability to carry one of the ship's kedge anchors slung underneath a boat that could be used for warping the ship.

There are wooden davits port and starboard on the ship's quarters for the cutters and these had the alternative name of 'quarter boats'. Immediately aft of these are iron curved davits for the gig and galley. At the stern, another davit is provided for the jolly boat.

On either side of the funnels are the two launches. The launch on the port side has the pinnace stowed inside. On the starboard side the second launch has the cutter gig and dinghy stowed inside. These boats were known as boom boats and were launched by rope purchase taken from the main topmast.

Deploying boats was a favourite exercise of inspecting admirals. The boats could move 12 officers and 125 men at almost immediate notice carrying 20lb guns and provisions and ammunition for a week.

The ship's carpenter was responsible for the upkeep of the boats. The larger boats were constructed of mahogany, while the smaller ones were of elm or pine. The pinnace and launch were double diagonal-planked and the remainder were clinker built, which consists of the upper planking overlapping the lower.

ABOVE Starboard launch with a 20lb gun that could be deployed in the boat. *(Author)*

ABOVE LEFT The jolly boat at the stern of *Warrior*. *(Author)*

Type	Length (ft)	Weight (tons)	Lifesaving	Oars	Sail	Armament
Launch × 2	42	8	140	18 DB	2 × standing lug and staysail	20-pounder
Pinnace × 1	32	4.5	70	14 DB	2 × standing lug and staysail	–
Cutter gig × 1	20	–	19	4/6 DB	Mainsail and staysail	–
Dinghy × 1	14	–	6	4 SB	Foresail and staysail	–
Cutter × 2	30	2	49	12 DB	Dipping lug and mizzen	Rocket launcher
Galley × 1	32	0.75	28	6 DB	2 × lugsails	–
Gig × 1	30	0.75	26	6 DB	2 × lugsails	–
Cutter (jolly boat)	18	–	12	4 SB	Foresail and mainsail	–

Note: DB = double-banked oars; SB = single-banked oars.

Chapter Three

Firepower

HMS *Warrior* was the ultimate deterrent of her time, with the largest and most powerful array of naval firepower ever to put to sea. No other warship in the world could challenge her.

OPPOSITE *Warrior*'s 'teeth' – the port-side gun deck. *(Author)*

The guns

Charles Dickens' words that *Warrior* has 'as terrible a row of incisor teeth that ever closed on a French frigate', was his description of the guns on the most powerful warship the world had ever seen. When *Warrior* was new – and as restored – she carried a main armament of 36 guns: 26 × 68-pounders and 10 × 110-pounders. The guns were identified by the weight of the shot they fired. In addition there were 4 × 40lb guns, 2 × 20lb guns and 1 × 12lb gun to be carried in the ship's boats with one further 6lb bronze gun for practice firings. *Warrior* has a single gun deck with 34 guns. Two of the 110lb guns are mounted on the upper deck as chase guns. No longer were there tiers of gun decks as had been the case on traditional wooden warships.

It is the weight of the shot that is the crucial factor as to why *Warrior* was so powerful. When the Admiralty Board saw the early proposal for *Warrior* they were taken aback at the low number of guns and felt she was seriously underarmed – especially as wooden three-deck warships then under construction carried up to 120 guns. *Warrior*'s complement of just 36 must have set alarm bells ringing.

The philosophy behind *Warrior*'s armament was to have a ship with guns that were more powerful than any other warship's; combined with the ship's speed it could close on to an enemy and engage while protected by the armour. This was the tactical philosophy for *Warrior* that Baldwin Walker had always advocated.

Prior to *Warrior*, the Royal Navy had standardised on the 32lb cannon for its first-rate battleships. This is the largest gun that can be seen today on HMS *Victory*. With simple mathematics of multiplying the weight of the cannonball by the number of cannon, you have the total weight of shot that a ship can fire. When this calculation is applied the result is astonishing.

	Shot weight (lb)	No. Guns	Total (lb)	Total (kg)	Comparison
Victory	32	30	960	436	960lb
Warrior	68	26	1,768	803	
Warrior	110	10	1,100	500	
Warrior	40	4	160	72	Total 3,028lb

The *Victory* had a total broadside weight of 960lb (436kg), while *Warrior* had the unprecedented weight of 3,028lb (1,375kg) broadside in comparison. That's three times that of *Victory*. You no longer counted guns: instead you counted the weight of shot the ship could fire. These figures had an impact not only in the physical sense but with the political and public perceptions that at last the Admiralty were doing something. With what today we would call public relations, the government made capital from the ship and her capabilities. The country as a whole was looking to *Warrior* as the answer to curbing France's naval ambitions.

The 68lb cannon

The 68lb (31kg) cannon was an artillery piece designed and used by the British armed forces from the mid-19th century. The cannon was a smoothbore muzzle-loading weapon produced in several weights. At this time all design, manufacture and purchasing of artillery was controlled by the British Army Ordnance Department which issued guns to the Royal Navy. The navy did not achieve autonomy over its own guns until the late 1890s.

Colonel William Dundas was the government's Inspector of Artillery between 1839 and 1852. He designed a 5.6-ton cannon in 1841, followed by a number of variants of which the most common was the 4.75 ton dating from 1846 – these were the cannon issued to HMS *Warrior*. With its gun carriage the total weight was 5.5 tons. It served with both the British Army's Royal Artillery and the Royal Navy and was used extensively in the Crimean War.

The cannon were cast by the Low Moor Iron Works in Bradford from 1847 and were relatively cheap to produce at £167 each. Over 2,000 were manufactured and gained a reputation as the finest smoothbore cannon ever made. The cannon were produced at a time when new rifled and breech-loading guns were beginning to make their mark on artillery. However the 68-pounder's reliability, simplicity and power meant that it was retained even on new warships such as HMS *Warrior*. Eventually breech-loaded rifled guns did make

smoothbore muzzle loaded cannon obsolete. This left the British forces with large surplus stocks of 68-pounders but these were given a new lease of life when fitted with a rifled sleeve to enable shells to be used. The 68-pounder was used extensively in British coastal defences that were constructed during the 1850s as part of the response to the French threat. They were notably employed at forts like Gomer, Elson and Nelson which defended Portsmouth, and Victoria, Albert and Freshwater Redoubt defending the Needles Passage. The forts were known as 'Palmerston's Follies' after the Prime Minister who had ordered them to be built. Just like *Warrior*, they never fired a shot in anger. The rifled versions of these guns remained in service throughout the First World War in shore fortifications and were not declared obsolete until 1921.

ABOVE One of the 26 68lb cannon on board *Warrior*. *(Author)*

BELOW A 68lb cannon on its gun carriage. *(Gary Cook)*

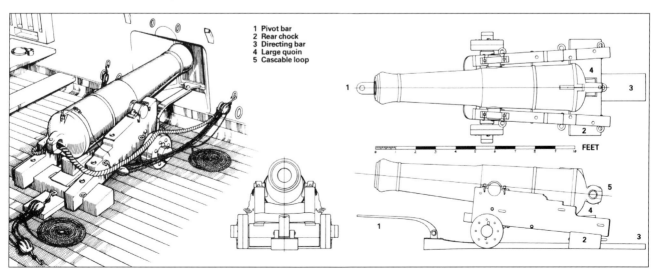

1 Pivot bar
2 Rear chock
3 Directing bar
4 Large quoin
5 Cascable loop

What it could do

The gun was a traditional muzzleloader which had to be loaded from the end of the barrel. Before it could be loaded, the gun was pulled back about 6ft (2m) from the ship's side to give access to the bore of the barrel, which was cleaned with a sponge. On *Warrior* there are two types of sponge; one is attached to the long pole stowed in the deck head above the guns. It would be pushed out through the open gunport and then back into the cannon muzzle. Having the gunport open was a potential hazard if the ship was in action so an alternative was to use the flexible sponge that hung above each gun on the end of a long rope; by twisting the rope it could be pushed down the barrel while allowing the gunport to remain shut.

The powder charge came from either of the two magazines in a leather satchel, which was delivered to the gun and rammed down into the breech. There were three weights of charge that could be used: 8lb, 12lb and 16lb. A wad made from shredded rope was pushed down as a seal, again using either the pole rammer or flexible rammer. This was followed by the shot. Although the cannon's barrel bore was slightly over 8in (20.6cm), both shot and shells were just under 8in (20.1cm) in diameter. This allowed for what was called a windage gap of 0.1in (0.25cm) around the projectile; enough to aid the loading process, but not enough to seriously diffuse the exploding powder.

RIGHT Gunpowder charges were delivered to each gun in the leather satchel. Three weights of powder charge could be used on the 68lb cannon. *(Author)*

To prevent the embarrassment of the shot rolling out of the cannon if the ship rolled in a seaway, a rope grommet was pushed into the bore to hold the shot and charge in place. The gun was then pulled forward to the firing position using the crew's muscle power with the block and rope tackle on either side of the gun. Once in position the tackle was unhooked. The barrel did not protrude out of the ship but stayed parallel to the ship's side. The gun was primed using a metal spike inserted through the firing vent that is located towards the rear of the gun. This pierced the charge bag, then a goose quill filled with gunpowder was inserted down the vent and a percussion cap fitted on to the firing hammer that, when struck, ignited the charge and forced the shot out of the barrel. There was an alternative friction firing mechanism that was used on some of the guns.

On firing, the gun and carriage recoiled back restrained by the breeching tackle, the large diameter rope passing through the large eye plate at the rear of the gun known as a cascabel. The cannon had a range of approximately 3,000yd (2,700m). At its maximum elevation of 15 degrees it had a maximum range of 3,600yd (3,310m), a distance that the shot would cover in 15 seconds and a muzzle velocity of 1,579ft/s (481m/s).

As with all smoothbore weapons, accuracy was not comparable to rifled guns. Despite the impressive range there was only a 10% possibility of hitting the target at such extreme distances. Reducing the range increased the accuracy enormously; for instance, at 600yd the cannon had 80% accuracy.

The cannon could fire a range of projectiles, solid shot (cannonball), explosive shells, grapeshot, case shot and 'Martin's Liquid Iron Shell'. The shot (cannon ball) was manufactured as cast iron shot weighing 67lb (30kg) or wrought iron shot weighing 72lb (33kg). One shot had the destructive power against a wooden warship equivalent to five of HMS *Victory*'s 32lb guns.

The explosive shells were filled with 4lb (1.8kg) of gunpowder and fitted with a timed fuse that had been designed by Captain Edward Boxer in 1849 and so was consequently known as the Boxer fuse. It was a paper cone calibrated in seconds along the

side. By perforating one of the numbers, the fuse was ignited on firing the gun and the shell exploded at the selected time. The gun captain still, however, had to gauge the best length of fuse for the range at which the ship was firing. To prevent the shell exploding in the barrel it was fitted with a sabot (wooden base) to ensure the fuse faced away from the charge.

Grapeshot comprises 15 × 3lb balls in three layers held together by wooden discs and an iron rod which broke open on firing. These were used against small ships or open boats and were predominantly anti-personnel rounds. Case shot, meanwhile, is a red cylinder containing musket balls and was also used as an anti-personnel round.

The 'Martin's Liquid Iron Shell', as its name suggests, is a shell filled with molten metal and was the equivalent of red-hot shot used in

earlier ships and modern napalm. In *Warrior's* forward stoke hold is a metal furnace in which the liquid iron was produced. The crew had a maximum of 4 minutes to bring the extremely hot shot from the boiler room and into the gun ready to fire otherwise the metal would begin to solidify. It was hazardous to use and the ship's

ABOVE Cleared for action: the mess table and benches have been lashed into the deck head, giving a wide clear space between the guns. *(Author)*

RIGHT The pivot bar. *(Author)*

RIGHT Siting marks on the 68lb cannon. *(Author)*

RIGHT Wheeled handspike.

records show only one occasion, in November 1862, when it was fired.

Each gun had a crew of 18 and in action only one side of the ship's guns would be manned. In the unlikely event of having to engage an enemy on both sides, the crews could be reduced to nine men to enable the opposite guns to be brought into action, though obviously this meant that the rate of fire would drop.

The majority of the gun crew were there to provide the muscle power needed to move the gun. 'Clearing for action' had to be achieved in 4 minutes, the mess tables and benches were lashed up into the deck head which left a clear space between the guns. Foresight had been shown by Walker and Watts when designing *Warrior* – they decided to increase the spacing between the guns from 8ft to an unprecedented 15ft, giving far more room for the guns' crews. This allowed the guns to be traversed so they could fire at an angle to the gunport. The pivot point of the gun, known as the fighting bolt, is on the gunport frame and using a handspike and plenty of men on the rear of the gun it could be pivoted to fire at an angle. The marks in the deck head of an arrow or heart indicate the angles of 17 degrees and 28 degrees that the guns could be moved to.

The Royal Navy was famous for having the best-trained crews and they practised and practised until they were proficient in gunnery drill. Each gun crew was expected to achieve a shot every 55 seconds. Similar to modern-day F1 motor racing pit crews, every man had a co-ordinated and specific task. Reality was slightly different when at sea as the roll of the ship had to be taken into account. Immediately above the gun is a hook on which a pendulum hung and the gun fired when the pendulum was vertical. This produced a rate of fire of about 90 seconds in a rolling sea.

The 18 men on a gun were numbered and assigned specific duty:

■ Nos 1 and 2 were the captain and the second captain of the gun
■ Nos 3 and 4 were loaders and spongers
■ Nos 5 and 6 were assistants to 3 and 4
■ The remaining crew were known as handspikesmen, assistant and roller handspikesmen, rearmen, stationary powderman and powderman.

The cannon was fitted with sights. A vertical arrow foresight was fitted halfway along the barrel and an advanced hexagonal adjustable sight was positioned at the rear. The rear sight was calibrated on each face with a designation of what type of shot was being used and different weight of gunpowder charge (for example, 'S' was the definition for 'shell' being case, grape, explosive or liquid metal shell).

Mark	Charge (lb)	Description
R	8lb	Reduced charge
F	12lb	Full charge
D	16lb	Distance
SR	8lb	Shell reduced charge
SF	12lb	Shell full charge
SD	16lb	Shell distance
OMEGA		

Having pushed the rear of the gun around to align on the target, the elevation was adjusted by moving the sight up or down until bearing on the target and then reading off the number from the column marked 'Omega'. The large calibrated wedge under the cannon, known as the quoin, was then driven in or out to correspond to the sight number. The gun was now sighted in both elevation and azimuth.

A percussion cap was fitted to the

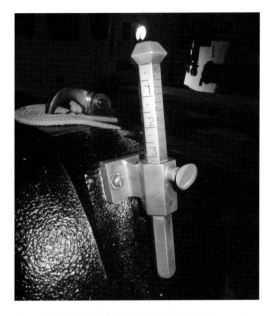

LEFT Gunsight.

LEFT The wedge under each cannon was known as a quoin and was moved to the number that had been read off the gunsight to adjust the elevation. *(Author)*

FAR LEFT The indicator board shows the magazine crew which size charge to send up to the gun deck. *(Author)*

LEFT A mop was used with water from the bucket to cool the gun barrel externally. The rope grommets were pushed down the bore of the cannon to prevent the shot from rolling out prior to firing. *(Author)*

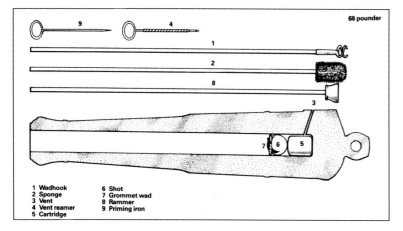

68 pounder

1 Wadhook
2 Sponge
3 Vent
4 Vent reamer
5 Cartridge
6 Shot
7 Grommet wad
8 Rammer
9 Priming iron

ABOVE The 68-pounder gun. *(Gary Cook)*

detonating hammer, or trigger, and when the gun captain was satisfied that the gun was correctly aimed on to the target, the gun would be fired by jerking the firing lanyard. As the gun recoiled, the slack of the restraining tackle was taken up and the now hot gun was immediately sponged out and the loading process commenced again.

RIGHT The Armstrong 110lb breech-loading rifled gun. *(Author)*

At each end of the armoured gun deck there was an indicator board where the arrow marker could be moved. This, in turn, moved a corresponding indicator in the magazine to inform the men which size charge to send up to the gun deck. One mark was for a charge bag filled with sawdust and was used for exercising the gun crews in firing drill.

The records show that this practice of selecting the size of powder charge was not always followed and full weight charges became the norm as speed of loading and power of the shot was the predominant requirement in any action.

110lb Armstrong gun

It was recognised that considerable developments in rifled guns were occurring and the performance of these new weapons was promising in being able to provide not only increased destructive power but also better accuracy at greater ranges. There were claims that these guns would also penetrate armoured ships. This was a radical new direction for artillery and it could not be ignored.

William Armstrong had produced a series of rifled breech-loading (RBL) field guns for the army from 1855 firing 6lb, 9lb and 12lb shot and these were viewed as very successful. They were easy to load, easy to manoeuvre and accurate. The Select Committee on Ordnance pressed Armstrong to build larger guns driven by the continued speculations and rumours that

RIGHT 110-pounder Armstrong gun on rear chock carriage. *(Gary Cook)*

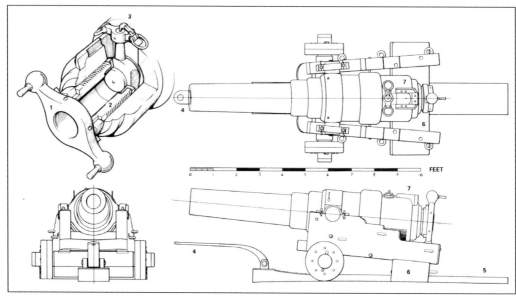

FEET

the French were installing large rifled guns on their new warships.

Armstrong had reservations about building larger guns and protested that the breech mechanism was unsuited for a heavier weapon, though eventually the commercial gains his company would enjoy drove him to produce a 20lb gun, followed by a 40lb and then the 100lb gun which developed into the 110-pounder.

The Armstrong guns used a screw breech mechanism that involved a heavy block inserted into a vertical slot in the barrel behind the loading chamber, with a large hollow screw behind it which was manually tightened against the block after loading. A metal cup on the block, together with the pressure of the screw behind it, provided obturation and sealed the breech to prevent escape of gases rearward on firing. The sliding block was known as the 'vent-piece' as the vent tube was inserted through it to fire the gun. The gun's rifling used the 'polygroove' system, giving the bore of the gun 38 grooves along its length with a twist of one turn per 38 calibres. (A calibre is a measure of the diameter of the gun multiplied by 38 – therefore 8in diameter × 38 = 304in.)

To load the gun, the hollow screw was backed off, removing pressure from the vent-piece which could be lifted out. The shell was inserted through the hollow screw and rammed home into the bore, then the powder cartridge was likewise inserted through the screw into the chamber. The vent-piece was then reinserted, the screw was tightened, a firing tube (a goose quill filled with gunpowder) was inserted in the top of the vent-piece, a percussion cap fitted on to the firing tube and the gun was fired.

The shells had a thin lead coating which made them fractionally larger than the gun's bore, which engaged with the gun's rifling grooves to impart axial spin to the shell. This spin enabled the gun to achieve greater range and accuracy than existing smoothbore muzzle-loaders.

On top of each powder cartridge was a 'lubricator' consisting of tallow and linseed oil between two tin plates, backed by a felt wad coated with beeswax and finally by millboard. As the lubricator followed the shell down the bore, the lubricant was squeezed out between the tin plates and the wad behind it cleaned out

LEFT The 'polygroove' rifling of the Armstrong gun. *(Author)*

any lead deposits left from the shell coating, leaving the bore clean for the next round.

Armstrong's early models in test firings had proved to be reliable and accurate over extended ranges when compared to smoothbore cannon. The gun could fire shells fitted with either percussion fuse or timed fuses with good results. The navy became highly enthusiastic, to the extent of advocating that all future guns should adopt the Armstrong

BELOW The 110-pounder gun. *(Gary Cook)*

BOTTOM Drawing of the lubricator and charge. The lubricator was designed to clean lead off the rifling but in service it was not that successful. *(WPT)*

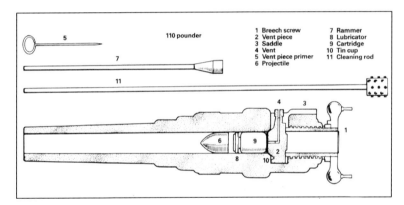

110 pounder

1 Breech screw
2 Vent piece
3 Saddle
4 Vent
5 Vent piece primer
6 Projectile
7 Rammer
8 Lubricator
9 Cartridge
10 Tin cup
11 Cleaning rod

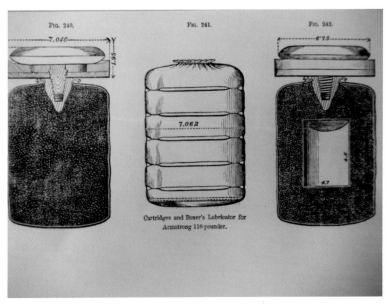

Fig. 240.   Fig. 241.   Fig. 242.

Cartridges and Boxer's Lubricator for Armstrong 110-pounder.

design. They also wanted *Warrior* to be entirely equipped with rifled breech-loading guns, but Armstrong was unable to manufacture his larger gun in sufficient numbers to meet her planned commissioning date.

However, reality did not keep pace with the enthusiasm. In January 1859 an early version of a 40lb Armstrong gun was test fired at 4½in armour fitted to the floating armoured gun battery *Trusty*. Contrary to the high expectation of seeing total destruction, the shot did not penetrate the armour even when the range was reduced to less than 50yd. Somewhat surprisingly, the Ordnance Board dismissed the results and the 40lb gun was accepted for service without further trials.

BELOW **A 110lb Armstrong secured in its seagoing stowage. All the guns would be secured in this manner when seas became rough.**
(Author)

Another set of trials was conducted in late 1859 with Armstrong's next gun, an 80-pounder which claimed to fire armour-piercing rounds. Once more *Trusty* was the target and again the results were disappointing. This was compounded by the 68lb muzzle-loading smoothbore cannon achieving better results when fired at the same target.

Despite these repeated failures, and with a degree of incredulity when one looks back, the War Office and Admiralty accepted the 80lb gun and placed production orders for the 110lb version that had not even been built.

Politics were at the heart of this high-risk decision and William Armstrong was a master of influence. With *La Gloire* completed and supposedly equipped with the latest rifled guns, the French threat was now a reality and to have fitted only smoothbore cannon to *Warrior* would have been an open admission to an already critical British press, the British people and the French that Britain had been unable to match their technology. The guns ordered may not have been very good but they were saving the nation, and quite a few admirals and politicians, from humiliation. Indeed in 1863 the Ordnance Committee admitted that the urgency of acquiring this gun had not allowed sufficient time '. . . for maturing the design previous to its manufacture'.

When the 110lb gun was manufactured it became apparent in test firings that it could not fire the same 16lb charge that the cannon used without the risk of the breech vent-piece blowing out. This weakness had forced Armstrong to reduce the propellant charge to 14lb and by 1865 to 11lb. This reduced the muzzle velocity. The cannon, with its 16lb charge, achieved a velocity of 1,750ft/s but the reduced 11lb charge produced only 1,100ft/s and lacked the penetrating power needed to defeat armour.

The aspiration of full conversion to the Armstrong gun was reined in. The navy accepted, as a compromise, a mix of the 68lb cannon with a proportion of Armstrong guns. The cannon were for attacking armoured ships at close range and the rifled guns would be for long-range attacks on wooden warships.

Further gunnery trials were undertaken in October 1861, after *Warrior* had been commissioned, at the Shoeburyness artillery ranges in Essex when a 20ft mock-up of *Warrior*'s armoured belt was subjected to shots from both the 110lb gun and 68lb cannon. Neither weapon had any effect on the armour, with the 68lb cannon again proving the better gun.

Once in service, the Armstrong gun began to exhibit a number of worrying problems. The lead coating on the shells came off, especially if the gun was already hot from previous firings, filling the rifling grooves and turning it from a rifled gun into a smoothbore cannon. The lubricator designed to overcome this problem did not work that well in removing the residues. Consequently shells fired from a 'fouled' gun would topple and lose their ballistic accuracy. Manually cleaning the lead from the gun was difficult and time-consuming.

A further problem developed that after firing a number of shots the gun became hot, expanding the metal which caused the vent-piece to jam and thus preventing the gun being reloaded.

Cannon up to this time had always been manufactured by casting, but the Armstrong gun was produced by shrinking four wrought iron coils of diminishing length on to an inner barrel. After a few firings, hairline cracks began to appear on the larger of the iron coils. Though it was claimed these had no effect on the safety of the gun – and there is no evidence that one ever burst – it did not give the gun crews a great deal of confidence when they were the ones having to fire it.

Though *Warrior* never fired a shot in anger, the gun reached its nadir when used in other ships at the Bombardment of Kagoshima in 1863, an engagement brought about by the murder of a British trader in the Japanese town. Seven ships of the China Squadron equipped

with a number of Armstrong 110lb guns were dispatched to collect a financial indemnity imposed by the Japanese government on the town. However, on arrival the fortified town opened fire on the ships, killing a number of British sailors. The ships bombarded the town in retaliation. In the heat of the action over 300 rounds were fired by the Armstrong guns with 28 malfunctions, the most serious of which were attributed to not tightening the breech screw and the vent-piece blowing out, with the blast knocking over the gun crew.

With this catalogue of problems continuing to accumulate, by 1864 the 110lb Armstrong guns were withdrawn and the Army Ordnance Board and Admiralty lost confidence in large breech-

ABOVE The 110lb gun and its ammunition. *(Author)*

BELOW 8in guns were installed on *Warrior* later in her active life. This photograph is of HMS *Northumberland*, one of the five-masted ships built in 1866. *(NMRN)*

40lb Armstrong gun

There are four 40lb RBL Armstrong guns on board *Warrior*. They were manufactured from 1859 to 1863 and did not exhibit the problems of the larger 110lb version, despite having the same screw breech design, the reason being that it used a smaller 5lb charge. It survived in the armed forces until 1877, by which time over a thousand had been built. Primarily used by the army as a mobile artillery piece, they had a limited offensive capability on board ships of *Warrior*'s size where they were used as little more than saluting guns. They were effective, though, when installed on smaller ships such as frigates, sloops and gun boats. As their name suggests, these guns fired a 40lb shell. The bore of the gun was 4.7in, the weapon weighed 1¾ tons and had a range of 4,500yd. The carriage that the gun was mounted on was a modified 32lb gun carriage. Despite its relative small size it still had a crew of 12 men.

20lb gun

There are two 20lb RBL guns of 16cwt on board. This was a shortened version of the land-based 25lb army model and has a stubby appearance. Its short barrel only allowed a muzzle velocity of 1,000ft/s using a 2½lb charge and a range of about 3,400yd (3,100m). Both guns were designated as 'boat guns' for carrying in the ship's launches to support a landing party ashore.

When on the ship, the gun was mounted on a carriage fitted with wheels, which was known as a truck. When taken ashore the gun was transferred to a field carriage and limber that carried the ammunition and powder. Though the gun could be fired by a crew of 5 when on board the ship, 18 men were assigned when it was used ashore. Their main function was to provide muscle power in pulling the gun and limber to where it was required.

12lb gun

There is one Armstrong breech-loading 3in 12-pounder 8cwt 'boat gun' on *Warrior* and this was the earliest and smallest of Armstrong's rifled breech-loaders.

loaded guns (though the 40lb gun was retained). In *Warrior*'s first refit of 1864–67 she was re-equipped with 7in and 8in rifled muzzle-loaded guns firing a shell with studs that engaged with rifling grooves in the barrel. This was known as the 'Woolwich' rifling system. Looking like large Coca-Cola bottles, these guns were heavy – the 8in version weighing an unwieldy 9 tons – but were very accurate though with a slow rate of fire compared to their predecessors.

1867	Shell	Guns	Total lb	Kg
7in RML	160lb	28	4,480	2,032
8in RML	174lb	4	696	316
20lb	20lb	4	80	36
			5,256	2,384

It was not until the 1880s that Britain returned to large breech-loaders using the Elswick cup and the French De Bange systems, both of which used the power of the gun's firing to achieve obturation.

RIGHT The 12lb gun was another boat gun, but it was used mainly for firing salutes. *(Author)*

6lb bronze gun

The small bronze gun on the starboard side is a 6lb muzzle-loading cannon. It was a practice gun used to train the gun captains in judging the optimum time to fire when the ship was rolling in a seaway. The Admiralty accountants were much happier with the cost of firing this small gun than a 68lb shot and 110lb shell.

Small arms

There are 350 Enfield .577in muzzle-loaded rifles on board *Warrior*. Its formal description is an Enfield 2 band Pattern 1858 Naval Rifle Musket and it fired a Minié bullet by a black powder charge and a percussion cap on the firing nipple. A competent rifleman could fire three rounds a minute and the rifle was accurate to about 500yd.

The somewhat confusing description of a 'rifled musket' was at the insistence of the army who procured and issued guns for the navy at this time. Despite Enfield advising that the barrel of their rifle could be shorter than the musket it was replacing, the army insisted the barrel to be of the same length as the musket. When two rows of infantry were firing, the second row required a gun of sufficient barrel length to deploy safely over the shoulders of the first row of men – hence the description 'rifled musket'.

The rifles stowed adjacent to the senior officers' cabins have white webbing. These were the rifles used by the Royal Marines, and have a 'Yataghan' bayonet. This was a Turkish word for the double-curved bayonet that was unique to the Royal Marines. It was sharp along one side and on a third of the reverse side. When fitted, the rifle could still fire.

The other rifles have brown webbing and are for the crew. They have a long cutlass bayonet that when fitted obscured the ramrod. Having fired one shot, the rifle could not be reloaded while the bayonet was fitted and the weapon became a long pike.

There are 70 1852 Model Colt Navy revolvers of .36in calibre on *Warrior*. It was one of the first

LEFT 6lb bronze cannon used by the gun captains to practise firing 'on the roll' as the ship swayed in a seaway. *(Author)*

BELOW Enfield .577in rifles. *(Author)*

RIGHT **Colt Navy .36in revolvers stowed on a 'crocus' rack on top of the mooring bitts.** *(Author)*

BELOW **Yataghan bayonet.** *(Author)*

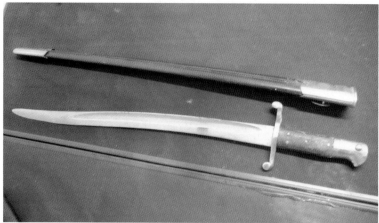

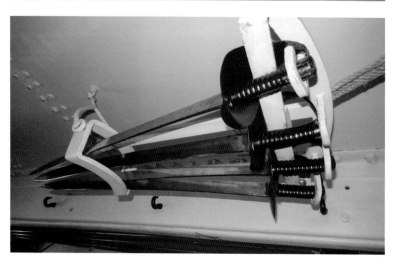

revolvers to adopt mass-production techniques whereby the individual parts were made en masse and then assembled on a production line. It was not a popular weapon as it proved difficult to load, taking between 15 and 20 minutes to pack black powder into each chamber of the cylinder and sealing with grease. However, it remained in service until 1880 when Enfield introduced an early cartridge revolver.

One of the revolver 'crocus' stowages on top of the mooring bitts is original to *Warrior*, having been found in the stoke hold when the restoration began.

Rifles, pistols and cutlasses were distributed throughout the ship (in fact 200 cutlasses and 50 7ft pikes as well as tomahawks). There remained a perception that boarding an enemy ship with its subsequent hand-to-hand fighting was still a viable tactic in future conflicts. But the technological advances had overtaken this outdated view. Instead, ships such as *Warrior*, with very large crews, could put ashore well-equipped landing parties supporting the army or civil authorities.

LEFT **Numerous cutlasses are stowed around the gun deck ready to repel boarders.** *(Author)*

Gunpowder (black powder)

Gunpowder is accredited with having been invented during the Tang Dynasty in China in the 9th century and was originally used more for fireworks than war. It was the warring Europeans from around the 14th century who developed it for guns. It is a mixture of sulphur, charcoal and potassium nitrate (saltpetre). The sulphur and charcoal act as fuels, while the saltpetre is an oxidiser. By the time of *Warrior* black power had become the term used in the military for gunpowder. Exactly why the description 'black powder' was adopted is obscure. It may stem from the fact that the first early smokeless powders were a light grey colour in comparison to the commonly dark black of the propellant it would ultimately supplant. Prior to that time black powder was simply called gunpowder, or in larger granulations, blasting powder or cannon powder.

Firing a 68lb cannon	Firing a 110lb Armstrong gun
■ Gun is cast loose, rear chock cast loose with roller handspike. Gun is run inboard such that the gun muzzle is 1ft clear of the gunport sill. ■ Gun is scoured with the wadhook and cleaned with the sponge. Vent is reamed with the vent reamer; to ensure there is a clear path, the sponge is passed up the barrel to ensure air is discharged from the vent. ■ Gun is now ready for loading.	■ Gun is cast loose and run in with roller handspike and train tackle, breech screw is slackened and vent-piece lifted out and placed on saddle. Captain of gun checks the bore is clear, vent-piece is replaced and breech screw tightened. Gun is run out with side tackles and roller handspike, vent is reamed and then cleared by firing a tube. Gun is now ready for loading and firing in the following order:
■ The vent is stopped with a vent plug.	■ Breech screw is slackened and vent-piece removed and placed on saddle and primed with vent-piece primer.
■ Cartridge, followed by shot and grommet wad are loaded into barrel to the extent of the loader's arm.	■ Projectile entered through breech screw and rammed home with rammer.
■ All three are rammed home using the rammer.	■ Lubricator is screwed on to cartridge and both inserted in breech lubricator first.
■ Priming iron is passed through the vent to prick the cartridge bag.	■ Tin cup is placed and the vent-piece is replaced and breech screw tightened.
■ Train tackle released, rear chock raised with roller handspike and gun run out with side tackle.	■ Gun is primed by placing tube in vent and firing lanyard put to half cock.
■ Gun is now primed by placing a goose quill tube in vent. Hammer (percussion firing) or firing lanyard (friction firing) to half cock.	■ While captain of gun holds the firing lanyard, gun is trained on to target by levering over the carriage with handspikes and hauling on side tackle to left or right.
■ With the captain of gun holding the trigger line or firing lanyard, the gun is pointed for training by levering over the rear of the carriage with handspikes and hauling on side tackles either right or left.	■ Gun is elevated to required elevation by adjusting the elevation screw assisted by the handspikes if necessary.
■ Gun is elevated to required elevation with handspikes and adjustment of the quoin (wedge) under the breech.	■ When captain of gun is satisfied the gun is correctly aimed, he gives order 'Ready'. No. 2 casts off half cock loop of firing lanyard and at the same time the side tackles are released.
■ When the captain of gun is satisfied the gun is correctly aimed, he gives the order 'Ready'. No. 2 fully cocks the detonating hammer or casts off the half cock loop of the firing lanyard and the side tackles are released.	■ When the sights are on target the captain of gun fires by jerking the lanyard. As soon as the gun has fired, the side tackles are manned and gun run out.
■ When the sights are on, the captain of gun fires by jerking the trigger line or firing lanyard. As the gun recoils, the slack of the training tackle is hauled in and held by the choking luff.	■ Vent-piece is removed, placed on the saddle. The tin cup is removed and the loading and firing sequence is repeated.
■ The captain of gun serves the vent, the gun is sponged out with the wet sponge to douse any smouldering residue.	

Chapter Four

Warrior in service

In twenty-three years of active service in the Royal Navy *Warrior* never fired a shot in anger, but she achieved what she was built to do: she protected the nation and her trade routes and she was the catalyst for the greatest advances in naval warship design the world had ever seen.

OPPOSITE Foredeck with the bow chaser gun. The opening in the bulwark leads to the crew's heads. *(Author)*

BELOW The reduced
bowsprit was intended
to save weight
forward. *(Author)*

BELOW RIGHT The
bowsprit support
crutch on the cable
deck. Note the gap
between the crutch
and the bowsprit
indicating the reduced
diameter of the smaller
and lighter bowsprit.
(Author)

First commission

In March 1862 *Warrior* and her crew were ready to begin her first commission as a warship of the Royal Navy, yet despite being ready for action she remained moored off Devonport. The dockyard at that time was not large enough to accommodate her. A mini refit was undertaken with the priority to alleviate the bow-down trim and prevent her plunging into heavy seas. Significant changes were necessary.

The jib boom was cut down by some 7ft and a shorter and smaller diameter bowsprit was installed. The anchors and their cat heads were moved aft by some distance. The two most forward 68lb cannon on the cable deck were moved into the master's and commander's cabins at the aft end of the ship. Sharing their cabin with a cannon was something they had to live with, despite their rank and status. The ship came first. These changes brought the ship to a level trim.

The crew's heads (toilets) were either side of the bowsprit and had been inaccessible when the ship was in any sort of active sea state. Cochrane had new cubicle heads built halfway along the bulwarks on the upper deck. They did not add to the appearance of *Warrior* but that was not

an issue a grateful crew would have worried about. These alterations took nearly three months to complete before *Warrior* joined the Channel Squadron. This she did in June 1862.

As an amusing aside, the Admiralty with their ultra-conservative Sea Lords were having difficulties over where to place *Warrior* and *Black Prince* in the order of battle as they needed to be 'rated', based on the number of guns and crew embarked. This was a system that had existed even before the *Mary Rose* back in the 16th century. The system had been standardised in 1677 by Samuel Pepys when he was Secretary to the Admiralty and was still being used in the 1860s; it was rigidly adhered to.

Rate	Guns	Crew
First	100+	850
Second	90–95	740
Third	64–85	500–700
Fourth	44–55	300–400
Fifth	32–40	250
Sixth	24–28	200
Brig/sloop	4–14	40–70

Applying these criteria, *Warrior* became a fifth-rate frigate and with her single deck of guns the inflexibility was such that she was called a 'frigate' despite being the most powerful warship ever built and possessing a crew of over 700 men. There is an anecdotal story of the Admiralty clerk responsible for manning getting very confused when he received a request for 700 men for a frigate, believing it to be a clerical error!

Without question she was a 'battleship'. But that was a new word that their Lordships struggled to accept. *Warrior* suffered from an identity crisis all her life and she went under frequent changes of type – eventually ending up with the rather neutral description of 'Her Majesty's Iron Warship "Warrior"'.

By 1862 two of the smaller iron warships, *Defence* and *Resistance*, had been commissioned and joined the Channel Squadron. These were two of the four iron ships that Baldwin Walker had reluctantly agreed to build, designed by Isaac Watts at half the size of *Warrior* (though considered only a quarter as powerful). Mounting only 20 guns and a speed at best of 11kts their contribution to the navy's strength was always questionable.

Warrior's early time in the Channel Squadron was a series of tactical trials and exercises with the other ships of the squadron. In particular, how she and the two other iron ships compared with the sailing properties of conventional wooden ships was investigated.

When *Warrior* joined the squadron it had deployed to Queenstown in Ireland (now Cobh

ABOVE HMS *Prince Consort* was not the most comfortable ship to serve on as she rolled alarmingly at sea. *(NMRN)*

permeated the Admiralty and the inability at that time for the naval dockyards to build large iron ships. *Prince Consort* was really a copy of the French *La Gloire* and she became notorious in naval service for rolling alarmingly.

Immediately after the launch the squadron sailed for Portsmouth where *Warrior* was docked to clean her hull. At the same time, further efforts were made to improve trim with the relocation of stores, 68lb shot, ropes and a number of water tanks further aft. This achieved a draught of 25ft at the bow and 27ft at the stern. *Warrior* now had ideal stern-down trim.

At this time Captain Cochrane had negotiated with the Admiralty to have a 'drying room' installed under the sail room equipped with a small coal-fired stove. This proved a great success in enabling hammocks and clothing to be dried irrespective of the weather and sea conditions. This was another first for *Warrior*.

Arthur Cochrane's practical mind continued to come up with improvements that could be made to *Warrior*, not least the construction of a small enclosed chart house on the forward bridge. Up to that time charts were hand drawn and had to be looked after with great care to protect them from the elements. The master kept the charts in his cabin, and there they remained, which imposed on him a long trek from his cabin to the bridge. Cochrane's chart house idea was the practical answer where the chart could be kept dry in the place where it was needed.

Warrior now put to sea for 'evolutions', which is the naval expression for practising drills, making sail, reefing sails, gunnery, navigating, manoeuvring and generally bringing the ship and her crew to the highest level of proficiency. Having achieved this high and demanding standard, you were expected to do it all again at night – bearing in mind that there were no lights, this was no mean feat.

The ship was inspected by their Lordships of the Admiralty predominantly to support their ongoing debates on what course future iron ships should take. *Warrior* had to perform various drills and exercises and the speed of making and furling sails, rather perversely, appeared to be the measure their Lordships used to ascertain the ship's capability.

in the Republic of Ireland). She encountered bad weather all the way but to everyone's satisfaction stayed reasonably dry, no longer shipping the large amounts of water over her bows.

After four days in Ireland the squadron sailed for Milford Haven in Wales to attend the launch of HMS *Prince Consort*, a 91-gun ship of the line from Pembroke Dock. Originally named *Triumph* it was changed to commemorate the late Prince Albert.

It was a strange event that a squadron of the most modern iron and steam-driven warships were assembled to witness the launching ceremony of a wooden warship. The *Prince Consort* was laid down in 1861 after *Warrior* had been ordered and was the first of five wooden warships that were being built, though they were iron clad and had steam propulsion. This epitomised the conservatism that still

BELOW HMS *Warrior*'s chart house installed on the forward bridge in 1862. *(WPT)*

In August 1862 *Warrior* was permitted to sail independently to the Scillies to further perfect her drills. On return she anchored in Portland, continuing to work up even when at anchor. While at Portland her sister ship *Black Prince* arrived, having been recently commissioned. The crew of *Warrior* manned the yards and cheered her in. There began an immediate friendly rivalry between the two vessels.

Both ships sailed together, repeating many of the exercises and evolutions that *Warrior* had already done. Much to the crew's delight *Warrior* bettered *Black Prince* in every respect. *Black Prince* never equalled *Warrior* for speed – she even burned more coal in not achieving *Warrior*'s speed. There were a number of other, admittedly minor, aspects in which *Warrior* proved superior. This was the penalty the Admiralty had to live with in having two 'identical' ships built in two different shipyards at either end of the country.

Shortly afterwards, both ships were assigned to the Mediterranean Squadron. Glamorous though this sounds, with images of calling at sun-kissed overseas ports, it really just signified joining the Mediterranean Fleet which was coming to the English Channel.

The Mediterranean Fleet was the only 'fleet' in the Royal Navy as the rest of the navy were in squadrons. To have command of the Mediterranean Fleet was one of the most prestigious commands a senior officer could achieve and a significant stepping-stone in reaching the very top in the navy. But *Warrior* and *Black Prince* had to make do with the second in command, Rear Admiral Sydney Dacres, who arrived in the 91-gun steam-driven wooden second-rate ship *Edgar*, accompanied by the 24-gun steam frigate *Liffey*. Though it was not acknowledged at the time, Dacres had been dispatched to review these new warships that were the talk of the country. He immediately ordered his small 'fleet' to Lisbon. When off Gibraltar, Dacres gave a further order for *Warrior* and *Black Prince* to steam together for one hour on six boilers. *Warrior* started about quarter of a mile behind *Black Prince* but by the end had overtaken her. Both ships passed Dacres; wooden warships '. . . as if they were at anchor'. At Gibraltar the crews were given 48 hours' leave and made fools of themselves ashore.

In November 1862 both ships returned to the Channel Squadron that was at Lisbon, now under Rear Admiral Smart. Smart was very much 'old school'. *Warrior* and her sister ship he referred to as 'hogs in armour' and he was intent on proving that the traditional sailing qualities of wooden walls and the tried-and-tested battle tactics were still better than these newfangled vessels. He nearly achieved his ambition, insisting that tactical manoeuvres were conducted in the 'Nelsonian' manner – a line of ships proceeding at the speed of the slowest sailing ship – which was something that *Warrior* and *Black Prince* were not designed to do.

The squadron was then sent to sea to measure the sailing qualities of the various ships and they headed into one of the worst squalls that *Warrior* had experienced. Sails were carried

ABOVE *Warrior* **anchored on the lower stretch of the River Tamar known as the Hamoaze, opposite Devonport Dockyard, Plymouth, 1862.** *(WPT)*

ABOVE A photograph of HMS *Black Prince* manning ship, as *Warrior* had done when escorting the royal yacht with Princess Alexandra and her father Prince Christian on board in March 1863.

away or split, gear swept overboard and, worst of all, the 100ft 15-ton main yard came down and broke. *Warrior* was forced to return to Lisbon for repairs.

Christmas 1862 was spent anchored in Madeira. Further exercises were conducted in the new year with Smart's conservative orders still restraining the iron ships from showing their power and capabilities. The squadron returned to Portsmouth in February 1863.

In March of that year the engagement of the

Prince of Wales (later Edward VII) to Princess Alexandra of Denmark was announced. The royal yacht *Victoria and Albert* was dispatched to bring the princess and her father, Crown Prince Christian, from Denmark to England. The seven ships of the Channel Squadron sailed as the escort to the yacht. They arrived in Denmark with due ceremony; gun salutes were fired and Captain Cochrane was invited on to the royal yacht to be presented to both the princess and her father. For the return journey, *Warrior* took station ahead of the yacht. When they anchored at Margate Roads 200 of *Warrior*'s sailors, dressed in their best white uniform, manned the yards and as the royal yacht came past gave three cheers.

Shortly afterwards the captain of the royal yacht informed Cochrane that the princess was impressed with *Warrior* and that she 'was much pleased'. Cochrane was delighted with the compliment and the performance of his ship and crew – so much so that he had a brass plate installed around the upper deck steering wheel engraved with the words 'Princess is much pleased'.

Warrior went to Devonport Dockyard for a maintenance period with the most significant task being to improve the steering gear. Cochrane's active mind recommended the installation of hydraulic-driven steering which he had received approval to install some

RIGHT HMS *Warrior* escorting the royal yacht *Victoria and Albert* in 1863, from a painting by S. Francis Smitheman. *(WPT)*

time previously. As this work disrupted his cabin, Cochrane went to London to report to the Ordnance Select Committee on the questionable performance of the Armstrong guns (see Chapter 3). Their conclusion was that the 110lb Armstrong gun should be removed from service, though they supported the advantages of rifled guns. In June *Warrior* was dry-docked in Portsmouth for ten days for another bottom-clean.

The First Sea Lord, the Duke of Somerset, was an astute politician who saw the value of showing off the navy; he ordered a round-Britain cruise of the Channel Squadron, calling at ports and coastal towns and opening the ships to the public. This was something unique as the public had only ever seen their ships from afar. The Admiralty was nervous of the idea and concerned more about what the crews might get up to ashore, though they conceded that it would be good for recruitment.

Back under the command of Admiral Dacres, the squadron embarked on the tour in July 1863, working their way around the country. At nearly all ports the ships had to anchor off as few could accommodate ships the size of *Warrior*. Off Sunderland over 190 boats delivered 13,000 people on and off *Warrior* in just 12 hours.

In the Firth of Forth, the crew made liberal use of Edinburgh. Shortly afterwards it was on to the Orkneys and then to Londonderry, followed by the Clyde at Greenock. At Liverpool, the mayor and corporation had arranged lunches, dinners and dances not just for the officers but for the crews as well – carefully selected from those who would behave themselves! – and the ship reciprocated the hospitality by welcoming 80,000 Liverpool residents on board.

While *Warrior* was on the River Mersey, Brunel's *Great Eastern* arrived. She was the largest ship in the world at that time, and made *Warrior* and *Black Prince* look insignificant by comparison. From Liverpool the tour continued across the Irish Sea to Dublin and finally it was home to Plymouth.

The cruise took three months in all, and by the end of it over a million people had visited the ships – 300,000 visiting *Warrior* alone. It was a huge success and the press were euphoric in

ABOVE Ship's wheels. The inscription commemorates the words of Princess Alexandra of Denmark in 1863 when she saw *Warrior*'s crew manning ship. Sadly, the original wheel and inscription was lost when the ship decommissioned. The restored inscription was unveiled by the current Princess Alexandra, who is *Warrior*'s patron. *(Author)*

their praise of the ships, the crews, the navy and – for once – the Admiralty and politicians.

In December 1863 *Warrior* and the squadron sailed for Madeira and, despite arriving on Christmas Eve, they conducted gunnery exercises. After Christmas it was on to Gibraltar then to Lisbon where Admiral Dacres conducted a detailed inspection of *Warrior* to his satisfaction. As a result there was another eventful run ashore for the crew.

In February the squadron went back to Portland where they remained until April. General Garibaldi of the Italian National Revival was a noteworthy visitor at this time, and gunnery demonstrations were carried out in his presence.

The next move was to Plymouth and into the newly built Queen's Dock. The hydraulic steering gear, fitted with the expectation of improving the helm, had been a failure and was subsequently taken out. A number of the Armstrong guns were removed and replaced with cannon.

In the final few months of her first commission *Warrior* carried out short cruises with the

BELOW HMS *Minotaur*, an enlarged version of HMS *Achilles*, was built in the same shipyard as *Warrior*. Ordered in 1863 and entering service 4 years later, *Minotaur* was described as a 'good steamer but unmanageable under sail'. *(NMRN)*

squadron to the West Country and to Ireland and one further trip to Gibraltar alone to bring back to England time-expired personnel, invalids and prisoners from the Mediterranean Fleet.

Admiral Dacres came aboard *Warrior* for her last voyage of the commission from Plymouth to Portsmouth, reaching Spithead in the Solent on 3 November. She moved to Sheer Jetty in the dockyard shortly afterwards. The ship was stripped of all equipment, boats, cutters, guns, stores and anchors. The crew had one final dinner aboard, then received their pay and left the ship.

Cochrane wrote to the Commander-in-Chief of the Royal Navy on 22 November 1864:

Sir,
I have the honour to inform that Her Majesty's ship under my command was this day paid off.
I have the honour to be, Sir,
Your obedient servant
Arthur A Cochrane.

Second commission

When *Warrior* completed her first commission, the Lords of the Admiralty and the politicians were still debating and, at times, openly arguing about the future of the navy. Many still believed that France's *La Gloire* was a superior vessel. The government had agreed to build a third Warrior and the order was given to Chatham Dockyard to build *Achilles* in their new dry dock. But significant changes were made; the ship was to be armoured from the bows to the stern.

News reached Britain that Napoleon III had ordered ten more ships in addition to the six already under construction. This would give France a two-to-one advantage over the Royal Navy. *No one* challenged the Royal Navy. Therefore action was necessary and the Admiralty Board immediately demanded ten new iron armoured ships and the conversion of ten wooden ships to armour plating. The government did not agree and, as always, a compromise was reached that five wooden ships would be armoured end to end and orders placed for three iron armoured ships larger than *Warrior*. These were the five-masted giants *Minotaur*, *Agincourt* and *Northumberland*. The order for *Minotaur*

LEFT USS *Monitor's* gun turret, showing impact damage after the Battle of Hampton Roads in 1862 during the American Civil War. *(Public domain)*

was placed with the Thames Iron Works and she was laid down on the same slipway immediately after *Warrior* had been launched. She was completed in 1867, just at the end of *Warrior's* first refit.

In 1862 an event on the other side of the Atlantic brought about a radical change in thought that marked the demise of the broadside armed warship. This was the battle during the American Civil War between the Union's *Monitor* and the Confederacy's *Merrimack*, two iron-plated ships armed with large guns that battered each other to a draw. Neither bettered the other. News of this caused a sensation in Britain and suggested that only iron-clad armoured ships would survive in a future conflict.

The use of a turret with two heavy guns on the American *Monitor* was also the precursor of adopting the same on Royal Navy ships. This idea had been advocated in Britain by the extremely influential Captain Cowper Coles, who persuaded the Admiralty in 1862 to order the first turret ship *Prince Albert* – a mastless coastal defence ship. She became the first turret ship in the Royal Navy to go to sea.

With iron eclipsing wooden ships, likewise the turret would see the broadside ship slip into obsolescence.

Warrior was in refit for two and a half years from 1864 to 1867. In that time the whole ship was overhauled, including the engines and boilers that were completely rebuilt. Changes were made to the watertight integrity with better subdivision of the hull with watertight doors added. But the biggest change was to her armament. The 68lb cannon and the Armstrong guns had gone, to be replaced with new 7in and 8in muzzle-loaded rifled guns (see Chapter 3). The old 40lb guns were also replaced with 20lb saluting guns.

BELOW HMS *Prince Albert* was the Royal Navy's first turret ship, ordered in 1862 and completed in 1866. Named after Queen Victoria's late husband, the ship remained in service for an unprecedented 33 years at the request of Her Majesty. *(NMRN)*

This was the largest rifled outfit of guns ever carried by a British ship. All were mounted on new iron carriages and slides. Eight of the 7in guns were mounted on the upper deck and the remainder on the gun deck. As a consequence, the magazines were rebuilt to handle the new ammunition.

Cochrane's chart house was moved from the forward bridge to the aft bridge and this proved to be a much more practical location. The ship also gained an impressive 42ft twin-screw steam-driven cutter.

As Warrior neared the end of her refit in early 1867, circumstances arose that required her presence. Not for war, but for prestige. The Sultan of Turkey and Khedive of Egypt were visiting Britain and the highlight was to be a fleet review with Queen Victoria.

As is inevitable, there were delays in the dockyard completing Warrior and in the navy finding sufficient crew to man her. Black Prince returned to Portsmouth to join the review and she transferred many of her crew to Warrior. She was quickly commissioned on 1 July with Captain Corbett in command and, with barely sufficient crews, both ships steamed from Portsmouth to Spithead.

The review must have been a unique sight. The 50 ships assembled clearly showed the transition that warships were making in the 19th century. A squadron of traditional three- and two-deck wooden warships, an iron-clad squadron, iron armoured warships and three small turret ships were all present. The review ended with a mock battle between the ships and the guns of the Spithead forts.

The ships dispersed at the completion of the review and Warrior was assigned again to the Channel Squadron. Captain Corbett left the ship on 24 July 1867 for another command, taking the crew with him. He passed the command to Captain Henry Boys who joined the ship the following day. Warrior was subsequently taken back into Portsmouth to receive a new crew.

In a navy run by strict protocol and regulations, it could be said that the 24 days of Captain Corbett's command was Warrior's second commission and that Captain Boys' tenure was the third. But history records that Boys' command was deemed the second commission.

It was still a prestigious command to be made captain of Warrior. Boys was considered by the Admiralty to be 'steady and able'.

The crew progressively arrived sufficient for the ship to undertake a measured-mile run. It was disappointing in only achieving just 12kts, rather than her previous best of 14kts. Various reasons were given, not least that the engineer claimed he had not been told it was a speed trial.

For over a month the ship anchored at Spithead, training and getting the routines in order. Some adjustment in crew numbers was needed to address specific issues, more cooks being one essential requirement.

She deployed to join the Channel Squadron at Queenstown (Cobh) in Ireland and then in company to Lisbon, enduring a storm en route. Warrior's steering again exhibited its customary problems when in heavy seas, with the stern lifting, the rudder losing all its effect, and, in addition, the ship broaching a number of times, turning 90 degrees to her course. The other ships of the squadron were newer vessels that showed their technical advantages over Warrior.

One exercise was a long, full power trial to find who was fastest. Achilles and Minotaur were deemed equal first with Warrior and Lord Warden equal second. In a memo of the Controller of the Navy are the words of Rear Admiral Fredrick Warden who was the Channel Squadron Commander on the fighting ability of his ships:

Report on Cruise of Channel Squadron
11 July 1867.

1st	Minotaur & Achilles	equal
2nd	Lord Warden & Warrior	equal
3rd	Bellerophon	
4th	Royal Oak & Prince Consort	equal
5th	Lord Clyde	

HMS Lord Clyde had the dubious distinction of rolling worse than HMS Prince Consort, both ships notorious during their naval service and probably why their gunnery skills never equalled that of their contemporaries.

In December, the squadron returned to Portsmouth for Christmas. But a threat to attack Queen Victoria while she was at Osborne House on the Isle of Wight by the Fenians –

who were attempting to overthrow British rule in Ireland – was a real possibility. *Warrior* and HMS *Irresistible* were anchored in Osborne Bay as guard ships. The crews, 'armed and ready', spent Christmas rowing the cutters around the bay and the Marines prepared to land if a signal was received from Osborne House. It probably wasn't the best Christmas the crew had ever had, but they *were* compensated for the two months *Warrior* was duty guard ship by being permitted 48 hours' leave in Portsmouth. Surprisingly there were few incidents recorded from these runs ashore.

Captain Boys was often invited to dine at Osborne House, usually with the household and on at least two occasions with Queen Victoria. There is an interesting note in his memoir, written later in life, where he states that at this time Her Majesty came aboard *Warrior* for an informal visit. Queen Victoria's diaries and the ship's logs (both meticulously kept) have no entry of such a visit, so whether or not it actually took place we may never know.

Warrior was dry-docked for ten days in March 1868. Back in the water, she ran the measured mile once more, this time achieving 14kts to the great joy of all on board. After this she went to Dublin as escort to the royal yacht for a visit by the Prince of Wales, and once this was over, she returned to Portland for more exercises and drills to bring the ship and her crew to their maximum efficiency.

In May 1868 there was a four-week Channel cruise and for ten days of that a flat calm sea prevailed, ideal for steam. However, at the insistence of the squadron commander, Rear Admiral Warden, sail was ordered to be used. It was not easy for a steam ship under sail to keep position in the squadron line when best speed was no more than 3–4kts in the light winds.

Warden was another admiral of the 'old school', who believed steam engines to be unreliable and preferred that sail should always prevail. There was some sympathy for his view from a logistical viewpoint as the liberal use of steam engines burned coal which cost money and would need replenishment – assuming the next port of call had coal.

In August 1868 the squadron sailed for a long cruise to Scotland and Northern Ireland. Now numbering some eight ships in two columns, they met squally weather on the night of 14 August and various signals from the flagship *Minotaur* became confused. As a result, *Warrior* was in collision with *Royal Oak* hitting on her quarter and destroying one of her cutters. *Warrior* lost her figurehead, jib boom and anchor stock. The wreckage took all night to clear. The cruise continued.

To the surprise of many, the Admiralty decided to court-martial *Warrior*'s captain for negligence and hazarding his and other ships. This action was possibly instigated by critical comments that had appeared in the press over the incident.

Henry Boys' court-martial was conducted on board *Royal Adelaide* at Plymouth. Boys defended himself as he considered it a nautical matter and that he was better availed of the subject than a solicitor. After four days he was acquitted, the charge unproven. There was rejoicing on *Warrior* and among the navy in general, Boys even receiving a congratulatory letter from the First Sea Lord.

What did emerge from the court-martial was the poor standard of signalling and communications between close-ordered ships at night. *Warrior* subsequently underwent repairs in Plymouth, including the figurehead.

In December 1868 a new commander of the Channel Squadron – Vice Admiral Sir Thomas Symonds – was appointed, and he hoisted his flag in *Warrior*. Immediately, the ship sailed south to Lisbon once more. Symonds was popular as he was a very experienced ship handler whom crews admired and had confidence in. However, he demanded strict discipline from them, as well as total efficiency. They conducted four cruises off Portugal and Symonds' realistic approach recognised that sail had had its day. He ordered all manoeuvring and exercises to be carried out under steam at greater speeds, despite the Admiralty's complaints over the cost of coal.

Early 1869 brought more drills, evolutions and gunnery practice, bringing the squadron to the peak of their ability and professionalism. The squadron returned to Portland in April that year and waiting for *Warrior* on her arrival were a set of somewhat unusual orders: she was to tow a floating dock to Bermuda.

RIGHT *Warrior*'s restored figurehead.

The formal description of *Warrior*'s figurehead is 'a full-length bearded figure with sword and buckler clad in Graeco-Roman armour'. It is 12ft (3.7m) tall, carved from pinewood and weighs 3 tons. It is an enlarged design of the figurehead that was fitted to the first (1781) HMS *Warrior*, a 74-gun wooden third-rate ship of the line.

James Hellyer and Sons was a firm of carvers and gilders whose business was based in Northumberland Wharf, Blackwall, and Portsmouth Dockyard and which had produced figureheads for the Royal Navy for generations. *Warrior*'s figurehead was carved by them in 1859 in the dockyard for the sum of £60. Incidentally, *Warrior* and her sister ship *Black Prince* were among the very last major warships to have traditional figureheads.

The practice of ships having ornamentation on their bows can be noted at several points in history. Viking vessels were among the first recorded to mount an aggressive figure at the bows to ward off evil spirits. Further back

RIGHT *Warrior*'s original figurehead at Victory Gate, Portsmouth, in 1883. *(WPT)*

in time, the Egyptians used holy birds on the prows of their ships, while the Phoenicians had horses representing speed. Roman ships, pre-dating *Warrior* by a few millennia, often mounted a carving of a centurion. Figureheads gained popularity with the galleons of the 16th century, when ship construction had developed such that a 'stemhead' was suitable for a figurehead. In a time when most of the population was illiterate, the figurehead was a way of indicating the ship's name.

HMS *Warrior* has had three figureheads in her lifetime. As described earlier, in 1868 *Warrior* collided with HMS *Royal Oak* when both ships were in the Channel Squadron; *Warrior*'s figurehead fell on to *Royal Oak*'s quarterdeck. An apocryphal story states that despite retrieving the body, *Royal Oak* claimed they could not find the head and consequently Hellyer's were commissioned to carve a replacement. It later emerged that the junior officers of *Royal Oak* had hidden the head and kept it as a trophy.

After *Warrior* decommissioned in 1883, the figurehead was removed from the ship and displayed in Portsmouth Dockyard adjacent to the Victory Gate. The entry gate is now used by visitors to the historic dockyard. It remained there until 1963 when it was transferred to Northwood in Middlesex where the new fleet headquarters were being established and had taken the name *Warrior*. Shortly after its arrival, however, it was found to be so rotten that it was scrapped.

The last figurehead is the one that looks down from the bows of the restored ship today, carved by Norman Gaches and Jack Whitehead at Wooton on the Isle of Wight. They started the task in 1981 using only photographs of the original as a guide. The work in progress was displayed at the 1982 London International Boat Show with the carvers still at work. Before it was finished in mid-1983, the figurehead appeared on the BBC children's television programme *Blue Peter*.

After completion it spent much of 1983–84 displayed at the Main Gate of Portsmouth Dockyard in the same location as its predecessor. Finally, it was taken to Hartlepool and mounted on the ship on 6 February 1985.

The Bermuda dry dock

Bermuda was a significant base for the Royal Navy as it was from there they could control the Caribbean and the eastern waters of the United States. However, there was no dry dock there, and the porosity of the island's limestone and sandstone geology prevented construction of a suitable one. To remedy this, Britain had built a massive floating dry dock measuring 381ft long × 123ft wide and over 74ft high. It had a huge U-shaped hull that could be partially submerged to allow a ship to be floated within it. The ballast tanks then pumped the water out, lifting the dock and bringing the ship clear of the water. Built by Campbell Johnson on the Thames, it was the largest floating dock in the world.

But it had to be towed to Bermuda. The plan was for HMSs *Northumberland* and *Agincourt* to tow the dock to Madeira where *Warrior* and *Black Prince* would take over. The voyage for all the ships would be under steam power, therefore coal consumption was critical. The paddle tug *Terrible* would assist. In Portsmouth *Warrior* landed all her upper deck guns. She took on board a large towing hawser and extra quantities of coal that filled

RIGHT **Arrival of the dock in Bermuda.** *(WPT)*

not only the bunkers but was stored in sacks around the ship.

She sailed with *Black Prince* in mid-June 1869 for Madeira, for what was to be the longest voyage of her career. Both ships arrived in the island without incident at the end of the month where coaling lighters were waiting to top-up each ship's bunkers.

The dry dock and her towing ships arrived shortly afterwards and the tow was transferred to *Warrior* and *Black Prince*. With the tug *Terrible* they set out in early July for Bermuda, 2,800 miles away. The dock was so large it had its own crew of 80 men, including a captain and surgeon, and it even followed naval routine including deck scrubbing and grog issue.

Sails were used on all the ships when the wind was right to ensure the most economic use of coal. The speed was no more than 4kts and *Terrible* was used to keep the ships and dock on course, owing to the fact that maintaining the correct heading was proving difficult.

The seas – fortunately – remained calm and after a 24-day voyage Bermuda was sighted on 28 July. The job was successfully done. Both ships sadly missed the celebrations of the Bermudian people who had gathered in large numbers to welcome them – they were too large to enter the port. They anchored, coaled and replenished stores and then sailed for home. The ships arrived at Spithead in late August to grateful thanks from the Admiralty for a job well done. Both of *Warrior*'s watches were given leave in Portsmouth, which they undoubtedly made the most of.

The popular Captain Henry Boys was promoted and left *Warrior* to another prestigious command. His replacement was Captain Frederick Stirling, but he was appointed for only six months.

The ship was dry-docked for a bottom-clean, the return of her upper deck guns and the installation of a complete new figurehead. Rejoining the Channel Squadron at Milford Haven, *Warrior* sailed for the regular deployment to Lisbon and then to the Canaries, Gibraltar, the Azores and back to Lisbon. Stirling was relieved by Captain Henry Glyn in March 1870.

Returning to Britain, *Warrior* met with the next generation of warships, the first-line turret ships *Monarch* and *Captain*, the latter being

very much the brainchild of Captain Cowper Coles. Though both turret ships, the pair were of very different designs. *Monarch* was a dry ship in heavy seas, whereas the *Captain* gained early notoriety of being a very 'wet ship'. Her freeboard (the distance from her main deck to the waterline) was extremely low, brought about by not controlling her weight as she was built. These ships were receiving the full focus of the press and public, meaning that *Warrior* and *Black Prince* were already moving into the lower echelons of warships.

Vice Admiral Sir Thomas Symonds took the Channel Squadron to sea again with the

ABOVE The dock in use during the late 19th century. *(Jonathan Falconer collection)*

BELOW A piece of Victorian publicity – a postcard of a painting by Thomas Dutton depicting *Warrior* in the Bermuda dry dock (with artistic licence, as *Warrior* never used the dock). *(WPT)*

new turret ships and headed into a double storm – one physical, the other political. The first was the bad weather from which the *Captain* emerged unscathed – Cowper Coles expounding his faith in his own design. The second involved Symonds being relieved of his command of the squadron due to his critical views of the Admiralty and their continued support for having sails, even in good steam ships. He accepted early retirement and left.

In June *Warrior* was in dock again for regular maintenance, to emerge six weeks later. At that time, she sailed for Gibraltar with the Channel Squadron and there joined with the Mediterranean Fleet for exercises. The combined force sailed into deteriorating weather. This developed into a storm in which *Warrior* lost two sails, but came off otherwise unscathed. HMS *Captain*, on the other hand, was not so fortunate. In the early hours of 7 September 1870 *Warrior* lost contact with HMS *Captain* in the poor weather.

During the storm she had capsized and sunk without other ships realising she was lost for some time. When it was discovered, the ships reversed course to search for her. Of a crew of nearly 500, they found just 18 survivors who had made it to the shore of Spain. This disaster represented more lives lost than at the Battle of Trafalgar. The exercise was immediately ended, *Warrior* and the squadron returning to Portland,

where she remained for three months. The news of the loss of the *Captain*, and of Cowper Coles who had gone down with the ship, brought shock to the nation as *Captain* had been looked upon as the new ship of the age.

Warrior was entering the last phase of her front-line service. She went to Portsmouth for Christmas leave for her crew and then back to Portland for another three months for static training of gunnery, boat drills and other exercises.

These harbour periods show in the ship's log an increase in disciplinary issues and a general decline of the rigid standards that had always been applied. The ship's chaplain was even put ashore for being drunk when conducting divine service.

In May 1871 *Warrior* sailed to Bearhaven in Ireland and then onward to Madeira. While en route the ship came upon the 1,100-ton iron merchant sailing ship *Hannibal* flying the distress flag. *Warrior*'s boats were dispatched to investigate and found that two of the crew had mutinied. They were brought back to *Warrior* and put in the cells before being landed in Madeira under the charge of the British Consulate.

The exercises continued with regularity off Gibraltar and there was a visit to Tangiers. In July *Warrior*, along with the Channel Squadron, sailed for Vigo in Spain. Off Ushant (an island

RIGHT The ill-fated **HMS *Captain*.** *(NMRN)*

off Brittany) they met the Mediterranean Fleet and the Reserve Squadron and when combined it was claimed to be the most powerful fleet of warships ever assembled in history. The group consisted of 17 iron warships, 4 large frigates and 2 corvettes. Exercises were conducted in perfect weather for two days and then the ships dispersed, after which *Warrior* sailed home to Portsmouth. She de-stored and on Friday 15 September 1871 she was paid off. Her second commission was at an end.

The table (right) provides a view of the distances run and the propulsion used during *Warrior*'s first two commissions.

Year	Steam (Nm)	Steam/ sail (Nm)	Sail (Nm)	Total (Nm)
1861	655	216	497	1,368
1862	1,712	3,657	1,288	6,657
1863	2,381	2,401	998	5,780
1864	2,225	3,043	564	5,832
Total First Commission	6,973	9,317	3,347	19,637
1867	556	1,785	–	2,341
1868	2,905	841	1,569	5,315
1869	3,111	4,985	5,132	13,228
Total Second Commission	6,572	7,611	6,701	20,884
Total 1861–71	13,545	16,928	10,048	40,521
%	33%	42%	25%	

Coastguard and Reserve Squadron

By the time *Warrior* left the Channel Squadron, the naval opposition put up by the French had dramatically changed. France never possessed the resources – either physically or technically – to keep pace with Britain.

La Gloire, France's great challenge to the Royal Navy and the catalyst for the very creation of *Warrior*, had been found to be deeply flawed. Only three ships of the class were built. *La Gloire* was the best of the trio as *Invincible* and *Normandie* had been built from unseasoned timbers that led to leaks and decay. They lasted less than a decade. The iron plates bolted to their sides raised their centres of gravity and that caused the ships to roll in even the mildest seas, such that the gunports were often immersed. Although they were claimed to have a speed of 13kts, in reality they barely reached 11kts.

France attempted to achieve standardisation with their warships by only having two or three different designs. The Royal Navy was the opposite, with a fleet that was technically diverse – in fact there were 30 major ships of over 20 different designs. Even ships considered to be the same were built by different shipyards and each had its own unique aspects, as was the case with *Warrior* and *Black Prince*. *Black Prince* never quite equalled *Warrior* in terms of performance. The support and maintenance of this varied fleet was a logistical nightmare.

However, the political map was changing dramatically at this time. Napoleon III had moved away from challenging Britain and had turned instead to the east and the emergence of Germany under the shrewd political leadership of Otto von Bismarck. This eventually culminated in war and France's humiliating defeat at the Battle of Sedan in 1870. Napoleon III was captured and the German army marched into France and occupied Paris for a short time before withdrawing, leaving a political vacuum and a broken nation. Napoleon, on his release, sought sanctuary from Queen Victoria. She assented and he lived his final years in Chislehurst, Kent, and is buried in Farnborough, Hampshire. Germany took up the mantle of being the new challenger to Britain.

Even though *Warrior* was moving to the second division she was still a capable warship. She was hard-hitting, had good speed and a hull that was still watertight and that would remain so throughout her life. Shortly after *Warrior*'s move into the Reserve Squadron, HMS *Devastation* was commissioned into the navy in 1873. As with *Warrior* a decade earlier, she represented another step change. She had no masts or sails, and instead relied on engines alone in her capability to steam over 5,500 nautical miles.

Her armament was in two large turrets on the upper deck, each housing two massive 12in guns that weighed 35 tons each. All this power was in the hands of a crew just half that of *Warrior*. There was no need for teams of sailors aloft setting and trimming sails.

Warrior entered Portsmouth Dockyard for a major refit that lasted from 1872 to 1875 in which she was totally overhauled. The significant change was a massive poop deck built at the stern of the upper deck to

accommodate an admiral. The forward capstan was converted to a steam-driven example.

She recommissioned on 1 April 1875 with Captain William Whyte in command. Her crew comprised just 350 men – just half of her original front-line complement of 700. *Warrior* was part of the 'First Reserve in the Coastguard Service and for Royal Naval Reserve duties'. This was quite a lengthy title, brought about by the traditional coastguard service having been amalgamated and run by the Admiralty. Her role was to defend the English coast from invasion by the French and to prevent smuggling. The latter task was considered ridiculous for ships such as *Warrior* to achieve.

Warrior conducted a full power trial in early May and she achieved 14.1kts, which was better than that achieved in her second commission. Based at Portland for some months, she took part in the summer cruise to Dublin Bay and then to Queenstown (Cobh). En route HMS *Iron Duke* rammed the 6,000-ton HMS *Vanguard* in dense fog. *Vanguard* sank but luckily there was no loss of life.

A potential threat from Russia caused the squadron to dispatch with some haste to Gibraltar to support the Channel Squadron when Russia and Turkey made hostile gestures to one another. That issue soon evaporated and

Warrior returned to Portland and to a sedentary period at anchor with the tedium of daily drills and maintenance routines.

Regatta week at Cowes on the Isle of Wight was her next duty, for which she was freshly painted and dressed overall with flags. Ceremonial duties continued throughout the week and her cutters were used to ferry royalty and dignitaries around the harbour. From Cowes she rejoined the Reserve Squadron and deployed to Spain.

In late 1877 war finally did break out between Russia and Turkey. Britain, fearing that the conflict could expand, mobilised her forces, including the Reserve Squadron that was brought up to strength. In early 1878 Captain Whyte was replaced by Captain Gordon Douglas. A squadron of 19 ships was gathered from the reserve ready to head to the Baltic. *Warrior* welcomed back Henry Boys, her second captain, who was now a rear admiral and second in command of the squadron. This gathering of Royal Navy strength had the desired effect and Russia backed down.

After this excitement had passed it was once again deployment to Ireland for *Warrior*. On returning in August 1878, the squadron assembled at Spithead for a review by Queen Victoria.

At this time naval strategy was changing from the old 'Nelsonian' practices of entire fleets of warships searching for the enemy. Instead the battleships were used for guarding British coasts and those of her Empire, while out at sea, fast cruisers and gunboats that probed the oceans scouting for enemy warships were needed.

Plans were discussed of converting *Warrior* and *Black Prince* to meet the cruiser requirement. They would have had new, more efficient and faster engines and would be armoured from stem to stern. The discussions dragged on well into 1880 but came to nothing.

There was yet another docking for *Warrior* in April 1880. Later that year there was a cruise to Spain and back to Portland for Christmas leave for the crew. There was a long list of defaulters afterwards from the various excursions into Weymouth. In January 1881 there was a further docking in Portsmouth and thence back to Portland.

The first half of the year consisted of more cruises, before returning to Portland where *Warrior* spent the rest of the year. Early in 1881 Captain Douglas had been succeeded by Captain Algernon Charles Fieschi Heneage. He was a man with a punctilious manner and an exaggerated view of his high position who looked down on those he regarded as socially inferior. He was an outrageous snob. His nickname was 'Pompo', an abbreviation of pompous. He was always immaculately dressed – having brought his valet on board – and he conducted white-glove inspections of the ship with the valet standing behind him with a tray of further white gloves. Heaven forbid if any dirt appeared on them.

The lowest in his mind were engineers. He even attempted to stop the crew saluting engineering officers, but naval regulations prevailed on that point and the saluting continued. At prayers he refused to wear his uniform as 'no Royal Navy officer should be seen kneeling to a higher deity'. His attitude did him little harm, however, and he rose to the rank of admiral, receiving a knighthood later in life. 'Pompo' lasted as captain for five months when *Warrior* exchanged crews with *Hercules* and Captain Samuel Townsend took command.

Dry-docked at Portsmouth, *Warrior* underwent a structural survey and was found to be in very good condition. In June 1881 she and the squadron visited Heligoland, Copenhagen and Kiel, where they were joined by a German ironclad squadron for a short exercise, then it was back to Britain. A cruise to Greenock and back to Portsmouth saw her encounter bad weather. It would turn out to be the last storm she would endure. *Warrior* lay at anchor at Spithead until the summer cruise of 1882 to Spain and Gibraltar, thereafter returning to Greenock.

In the November Captain Townsend was succeeded by Captain Edward Adeane, her last seagoing captain. She remained at Greenock well into the new year of 1883 when she was ordered south to drop anchor at Spithead on 11 May. Most of her crew left the ship and shortly after she entered Portsmouth to discharge coal and then head to South Railway Jetty, where her guns, equipment and stores were disembarked.

On 31 May 1883 a simple final line in the ship's log says: 'hauled down the pendant'.

LEFT Algernon Heneage, the most pompous captain who ever went to sea, was nicknamed 'Pompo'. This caricature of Heneage is by 'Spy' for *Vanity Fair* magazine in 1901. *(National Portrait Gallery)*

Chapter Five

Life on *Warrior*

A ship's crew was a community that required organisation and good working relationships to bring a vessel to the peak of its efficiency. *Warrior* represented a watershed in warship design and manning, where the technological advances of the Victorian age were beginning to take precedence over the muscle power relied upon in the age of fighting sail.

OPPOSITE The upper yardmen and midshipmen of HMS *Phoebe*, around 1860. Those of *Warrior* would have been very similar. Note the bare feet. *(NMRN)*

Crew

When one looks at the 18th- and 19th-century history of the Royal Navy, the hardships and brutality that the crews endured is what immediately comes to mind. From 1799 until 1815 Britain had been at war with Napoleon Bonaparte's France and the Royal Navy, with the largest fleet in its history, had needed men. Impressment, known colloquially as 'the press' or the 'press gang' – kidnapping fit men from the streets – was rife. The crews were effectively imprisoned on board ships and had to endure poor food and water, subsequent poor health and harsh discipline.

At the end of the Napoleonic Wars, and despite a huge reduction in the number of ships and their crews, Great Britain still retained the most powerful navy in the world – one that no other nation could rival. This was the backbone of Britain's economic, political and industrial

strength. Apart from small wars around the British Empire, the Royal Navy did not have a major fleet engagement again until Jutland in 1916.

The navy was not idle, though: they provided naval brigades ashore during the Crimean War; they hunted down slave ships; gave protection from piracy and ensured the nation and her Empire was secure. The public were justly proud. By this time the press gang was no more; you volunteered to join and there was no lack of recruits. To be in the navy was a prestigious career.

By the time of *Warrior*'s construction, the navy had become a far more systematic and professional organisation. Britain's industrial might was bringing to the fore talented individuals who may not have had high social standing but who were essential to the production and running of a modern navy and country.

The new steam ships were increasingly technical and a breed of men called engineers were emerging. Royal Naval Engineers became

BELOW The crew. This photograph may be of *Warrior*'s sister ship *Black Prince*, although very few pictures of either ship's full company exist. *(Walker Image Archive)*

a recognised department of the navy in 1833. In 1835 the ranks of first, second and third engineer were created. In 1847 an Inspector of Machinery was added along with the rank of Chief Engineer. Despite this, the navy clung to an increasingly outmoded class system that was elitist and socially influenced. Engineers were still looked upon as inferior.

In a similar vein the old adage of 'sail before steam' prevailed throughout the navy and dominated *Warrior*'s early years. Despite the power that steam brought in those first warships, it still did only one job – turning the propeller. It did nothing else to assist the crew, of whom nearly all were on board for one thing alone: muscle power.

Warrior's crew was 705 men and boys, but this was a fluctuating number as promotions, drafting and sickness moved men into and out of the ship in a continuous cycle. During her first four-year commission, 65 officers and 1,550 ratings passed through the ship. The crew list provides an overview:

Officers	36
Royal Marine Officers	3
Warrant Officers	3
Petty Officers	109
Seamen and boys (53 under 18 years)	344
Royal Marines SNCOs	6
Royal Marine Artillerymen	116
Chief Engineers	2
Engineers	10
Stokers and trimmers	76
Total	**705**

The table has condensed the crew into their respective groups. Every man on board had a specific job title associated with his duty. At the first commission these were:

■ **Officers**: Captain, commander, chaplain, master, naval instructor, paymaster, surgeon, second master, boatswain, carpenter, gunner, two assistant surgeons, master's assistant, assistant clerks, three assistant paymasters and clerks, five lieutenants, six naval cadets, ten sub-lieutenants and midshipmen.
■ **Warrant and Petty Officers**: Master at arms, chief gunner's mate, chief captain of the forecastle, chief quartermaster, chief carpenter's mate, seaman's schoolmaster, chief bandsman, ship's steward, chief boilermaker, ship's cook, founder.
■ **First Class Petty Officers**: Armourer, blacksmith, captain's coxswain, coxswain of the launch, captain of the hold, plumber, ropemaker, sailmaker, yeoman of signals. two captains of the afterguard, foretop, maintop, forecastle, carpenter's mates, caulkers, gunner's mates, four ship's corporals, seven bosun's mates, quartermasters.
■ **Second Class Petty Officers**: Caulker's mate, cooper, coxswain of the barge, coxswain of the pinnace, musician, two captains of the mast, captains of the mizzentop, coxswain of the cutter, sailmaker's mates, second captains of the forecastle, three signalmen, four second captains of the afterguard, second captains of the foretop and maintop.
■ **Seamen and boys**: One armourer, barber, blacksmith's mate, captain's cook and assistant, captain's servant, captain's steward, commander's servant, cooper's crew, lamp trimmer, second captain of the hold, ship's steward's assistant and boy, sick berth attendant, assistant and steward, tinsmith, wardroom cook's assistant, warrant officers' servant, warrant officers' cook, two butchers, cook's mates, engineers' servants, shoemakers, tailors, painters, wardroom/gunroom cooks and stewards, wardroom servants, gunroom servants, 3 sailmaker's crew, yeoman of storerooms, 5 shipwrights, 9 carpenter's crew, 15 bandsmen, 25 leading seamen, 44 boys first class, 8 boys second class, 199 able or ordinary seamen.
■ **Steam Department**: Two chief engineers, 10 assistant engineers, 20 leading stokers first class, 18 stokers/coal trimmers second class, 48 stokers/coal trimmers.
■ **Marines**: Captain, 2 lieutenants, drummers and buglers, 3 sergeants, corporals, 114 privates.

By 1860 getting into the navy, or as a parent getting your son into the service, was an ambition many aspired to. If you had a son and were not from the social elite, he would have had little or no education unless there was a church or charitable school close to where you lived. Ironically, orphaned children who were in the parish workhouses had compulsory schooling, so

'Bands of music' were on board ships as far back as the 16th and 17th century. The drum was the early method of signalling the change of watches and beating to quarters. Every ship had a drummer and a bugler. Bandsmen were not recognised by the Admiralty until 1842 when they created the rating of musician as a second class petty officer, permitting one per ship. His principal task was playing the violin when raising the anchor with the crew labouring at the capstan.

Five years later in 1847, the rating of bandsman was introduced and was open to any man who could play an instrument irrespective of his nationality. This attracted Germans, Spaniards, Italians and Maltese.

The Royal Marines band existed in the 19th century, having been formed in 1767, but they were based in the major Royal Dockyards such as Portsmouth, Plymouth and Chatham, and were not posted to ships.

Warrior's band was to an extent impromptu, relying on members of the crew with musical ability, with the emphasis placed more on string instruments than on wind. Until 1875 the captain and officers had to buy the instruments.

BELOW The ship's band and crew on *Black Prince*. *(Walker Image Archive)*

were perhaps better provided for educationally. Education did not become universal until 1880 with the introduction of the Education Act. Up to that time, even if a child had access to education, it provided little more than an ability to read and write, do simple sums and gain a good grounding in religion and the British Empire.

The Industrial Revolution, however, was creating huge social changes, as agriculture labour waned and factory work in the towns and cities, to which the population was migrating from rural areas, increased. There were no effective factory acts, no health and safety laws and life in manufacturing – in a mill or foundry or, worse, down a coal mine – was hazardous to health.

To men and boys, life in the Royal Navy, by contrast, seemed a good prospect, and by contrast with many jobs ashore, it was preferable. With his parents' agreement, a boy could join the navy and go into one of the training ships that were moored in naval ports such as Portsmouth, Plymouth and Chatham. He would complete one year's training and after reaching the age of 14 he would be posted to a ship where he would be rated 'boy 2nd class', and then at around 17 years rated 'boy 1st class'. At 18 he became a 'continuous service' man.

Until 1853 seamen had been recruited only for the duration of a ship's single commission, being discharged at the end. But in 1853 the Continuous Service Act was introduced that contracted a man into the navy for a fixed period, usually 10 years, with an option to extend by a further 10 years split into two five-year options. Whether he was on a ship or awaiting a posting in a depot ship, he was still paid and still in the navy.

The navy brought benefits, probably the most important of which was tenure of employment. There were not many jobs that guaranteed a regular income for 10 or 20 years. Furthermore, there were opportunities for promotion and the increased pay that came with it. Food and water was now good, there was access to medical care, a man was clothed (uniform being standardised in 1857) and generally well looked after. There was regular shore leave of between four and six weeks, especially if returning from foreign service. Travel passes were issued for those sailors not living in naval towns and a non-contributory pension was paid on retirement.

There were very few comparable lower-class jobs in civilian life that could match being in the navy. It was a coveted career.

It is interesting that the personnel records from the first commission show that the seamen in HMS *Warrior* had an average age of 27 years. The oldest man was 47. It would indicate that, with such a revolutionary and important ship, the very best and most experienced crew was assembled. Their average height was 5ft 6in, but there were a few who were 6ft tall.

The *Warrior* has been restored to her 1860 condition and it is therefore appropriate to provide brief biographies of a few of the significant people who first joined the ship.

Captain

To become the captain of a major warship, even today, is a prestigious and significant event in any officer's career. In the Victorian navy it was even more prestigious to receive a command. Captains of warships became public figures not dissimilar to today's sporting or television personalities.

During the Napoleonic Wars, ships' captains had powers that at times bordered on the extreme, working crews until they dropped from fatigue and meting out punishments that we would consider barbaric. Contrary to popular belief, the captain did not have the power to execute crewmen. Likewise, any punishment ordered had to be approved by the squadron commander and the Admiralty informed. By the time of *Warrior,* extreme disciplines remained on the statute books, but none of her captains would order such punishments in what was a peacetime navy. Flogging was still carried out and in 1862 over 1,000 men were flogged from a navy of 55,000, averaging 34 lashes each. Leave-breaking was the main reason, followed by drunkenness. Flogging was banned in 1881, although birching or caning remained a punishment up until the 1950s.

Within the navy the captain was known as 'second only unto God'. He was treated like royalty on his ship. When he came aboard it was to the shrill of the boatswain's pipe and the stamp of the marine sentry presenting arms. Officers on deck would remove their hats and any crewmen who were on the upper deck would be rigid at attention. His patronage was essential to any

LEFT Arthur Auckland Leopold Pedro Cochrane, *Warrior*'s first captain. *(WPT)*

aspiring junior officer or crewman who wanted to advance in the navy, so you did every ordered task to the best of your ability. The captain did not have the power to promote commissioned officers or warrant officers, but receiving a positive report from him on leaving a ship was extremely useful. Nobody could address the captain unless it was essential business.

When it came to selecting *Warrior*'s first captain the Sea Lords had a large number of potential people from whom they could select. There was no question that *Warrior* had the full attention of everybody in the country – royalty, the government, the general public and the press – as well as the rest of the navy with her imminent entry into service. She was going to change the world, but who would get to command?

There was a suggestion from an unnamed senior naval officer that the ship should first be taken to sea by a committee of suitably experienced officers to report on the general qualities of the vessel before appointing the command. However, this was not an idea that was taken up. For once the ultra-conservative Sea Lords made the right decision: they appointed Arthur Auckland Leopold Pedro Cochrane.

Cochrane received a letter from Captain John Moore, Secretary to the First Sea Lord, dated 31 May 1861:

The Duke of Somerset (the First Lord of the Admiralty) has desired me to inform you that it is his intention to appoint you to command the Warrior when she is ready for commission if it is agreeable to you. As it is desirable that whoever commands her should be present to superintend her fittings I believe you will be at once appointed additional to Fisgard for that purpose, but you had better come to the Admiralty as the Duke would like to see you on the subject.

Arthur Cochrane was the 37-year-old son of Admiral Thomas Cochrane, 10th Earl of Dundonald, who had become a legend during the Napoleonic Wars. Many claim that he had a reputation that could rival Nelson's. The elder Cochrane had been one of the most successful and daring captains, capturing many French and Spanish ships and earning the French soubriquet of 'Le Loup des Mers' (The Wolf of the Seas). He was very much an individual and never had the support of his peers as he loathed (and was outspoken about) the corruption both in the navy and in the government of the time. He and his crew were often denied prize money they were entitled to, often finding its way into

the pockets of others. Standing for Parliament as a radical he sought parliamentary reform in particular to rid the corrupt 'rotten boroughs' where wealthy individuals could buy their seat in Parliament. He was elected but in 1814 stood accused of a stock exchange fraud, lost his seat in Parliament and was dismissed from the navy. There is a story that says he had been tricked over a speculative investment and his dismissal was disproportionate – devised as a means to remove a man who challenged the unscrupulous and dishonest practices within Parliament and the navy. He went back to sea but in the capacity of what we would now call a 'mercenary', helping the rebel navies of Chile, Brazil and Peru in their wars of independence from Spain and Portugal. In those South American navies to this day his name is revered, with the Chilean navy always having a warship named *Almirante Cochrane*.

When his third son, Arthur, was born in 1824 he bestowed on him the additional Christian names Leopold Pedro in commemoration of his South American exploits. Arthur did not appreciate these unusual names and on joining the navy at the age of 13 was very particular in not using them.

In 1832 Thomas Cochrane had received a full pardon from the Crown and was reinstated into the Royal Navy with the rank of Rear Admiral of the Blue. He was made Commander of the North American and West Indies Station at the age of 70 and his son Arthur joined him as his flag lieutenant. He received several more promotions with the rank of Admiral of the Red and the honorary title of Rear Admiral of the United Kingdom. Queen Victoria and Prince Albert thought him a national hero and he was a regular attendee at court. The authors C.S. Forester and Patrick O'Brian are said to have based their naval characters of Hornblower and Jack Aubrey on Thomas Cochrane.

Thomas Cochrane died in October 1860 at the age of 84, just two months before *Warrior* was launched, and he never saw the ship or his son take command.

Arthur Cochrane inherited from his father natural leadership, good seamanship and (almost unique in senior officers of the day) an ability to understand and improve on the rapidly developing technicalities of modern ships. Equally

RIGHT **Thomas Cochrane, Lord Dundonald, was a naval legend in the Napoleonic Wars.** *(World History Archive/ TopFoto)*

he saw social improvements that could be made to better the lives of those serving on board.

In his 24 years of naval service prior to *Warrior*, Cochrane had already built a reputation that followed, though to a lesser extent, that of his father. There is no question that his name, along with his father's reputation, was advantageous for the younger Cochrane. Social pedigree was still a powerful influence in the promotion prospects of young officers.

Despite being in a supposed peacetime navy, he had seen action in the Middle East and the Crimean War, where he had his first command of the six-gun paddle sloop *Driver*. Promoted to captain, he went to the China Station in 1856 in command of the screw sloop *Niger* where he was wounded in a cutting-out action in the Pearl River against Chinese pirates. On his return to Britain he received the Order of the Bath and the China Medal.

Cochrane accepted his appointment to *Warrior*. There exists a rather concise letter which he wrote to his mother on 3 June 1861:

My Dearest Mother,
You will no doubt be very glad to hear that I am appointed to the Warrior, the finest and largest command in the British Navy. Hoping you are well,
your affectionate son
A A Cochrane.

During his command, Cochrane had numerous improvements made to the ship. Better heads (latrines) for the crew, a drying room for crew clothing and hammocks, a chart house on the forward bridge, changes in the type of ship's boats and attempts at improving the steering gear. He gave particular attention to the efficiency of the boilers as this was his forte, being an exponent of water-tube boilers rather than the smoke tubes that *Warrior* had.

He also made it his task to assist the paymasters in pursuing wages that had not been paid to seamen and ensuring they received campaign medals they were entitled to when on other ships. He was also concerned about bronchitis and tuberculosis that became rife among the crew and had suggested some fairly extreme solutions, including directing heated steam on to the gun deck where the crew lived.

A tradition in the Royal Navy that is still maintained to this day is that a captain cannot enter the ship's wardroom unless invited, and indeed Cochrane was invited to dine there on occasions. He respected tradition and would invite officers, especially junior ones such as cadets and midshipmen, to either breakfast or dine with him. This practice enabled him to get to know those younger men under his command.

Despite this he was no soft captain. When 700 men are first gathered together, there is inevitable friction between some that requires discipline and authority to ensure all work together. When a crewman named Harries and a boy called Morgan stole 2 guineas from another member of the crew and then deserted the ship, they were quickly arrested by the police and brought back to *Warrior*. Cochrane sentenced Harries to 48 lashes and Morgan to 36. The punishment was carried out before the crew on the upper deck. On completion, they were both discharged from the navy. In a rather perverse way, the crew appreciated this show of extreme punishment as it demonstrated that they had a captain who was fair but not afraid to impose discipline on those who broke the rules. He was greatly respected.

After *Warrior*, Cochrane took command of the training ship *Cumberland*. He was then appointed Superintendent of Sheerness Dockyard in 1869. In 1873 he became Commander-in-Chief Pacific Station and received a knighthood. Retiring in 1886 he was involved in managing the Trinidad Lake Asphalt Company. He never married and died in October 1905 aged 80.

Commander

The commander was the second in command and was responsible for the day-to-day running of the ship. *Warrior's* first commander was George Tryon. He and Cochrane knew each other well and they had both served under Cochrane's father when he was the commander of the North American Station; Tryon was the junior as a midshipman.

Tryon had begun his education at Eton College but halfway through informed his father he was joining the navy. He became an

RIGHT George Tryon, *Warrior*'s commander.
(Public domain)

outstanding young officer who had impressed all who served with him. Through his diligence he was appointed to the royal yacht *Victoria and Albert* in 1858, approved by no less a person than Her Majesty Queen Victoria. However, the job was a sinecure with no actual work to do, though it was useful for making contacts among the most influential people. This occurred in October 1860 when he was promoted and appointed commander of *Warrior*.

By the time of his appointment he had become an imposing figure, tall and of substantial build, with a massive black beard. He was intelligent, logical and forthright and everyone went in fear of him. It was Tryon's unquestionable authority and his knowing superiority over his contemporaries that brought a tragic end to his life. He had reached the rank of Vice-Admiral and in command of the Mediterranean Fleet when he gave the fateful order to change course and his flagship HMS *Victoria* was rammed by HMS *Camperdown*. The *Victoria* sank, taking Tyron and 358 of the crew with her. Those around him knew he had given an incorrect order but, such was his authority, nobody dared challenge it.

Master

The master was the ship's navigator. The title was a throwback to the sailing ship era where he would have been known as the sailing master. A sailing master would have consummate skills in navigation, coupled with knowing the characteristics of his ship from the set of the sail to the ballast and stability – knowledge that was only gained through long service at sea, and therefore his background was as a seaman and not the rarified officer class. However, it was recognised that after the captain, the master was the next most essential man on board. In 1808 he crossed that seemingly impossible social barrier of being allowed to enter and eat in the wardroom.

In 1843 the master became a commissioned officer but despite this seemingly privileged

BLAKEY'S STADIOMETER

The stadiometer is a rangefinder that determines how far away another ship is. It looks like a telescope, but has no lenses. On the barrel is an inset panel that holds a conversion table of tube length against height. There are two draw tubes with a scale of length inscribed along them. Looking through the eyepiece one would see two pairs of parallel wires at right angles to each other. The navigator would line up the top and bottom of the ship observed with the two parallel wires and then read off the length from the scale inscribed on the draw tubes. They would then use the table in the inset panel to work out the distance, based on the estimated height of the ship observed. Only two stadiometers are believed to have survived, one in the National Maritime Museum at Greenwich and the other on board HMS *Warrior*.

RIGHT The master's speaking trumpet and stadiometer.
(Author)

status he was still looked upon as a warrant officer. As a result, from 1840 to 1860 the number of masters in the Royal Navy halved, falling from 140 to fewer than 75, because their pay was lower than that of junior officers. With the rise of Britain's overseas trade there were far better paid berths for these experienced men in merchant ships. To rectify this, in the early 1860s the Admiralty began to phase out the title 'master' to be replaced by staff commander or officer. In 1867 the Navigating Branch, with appropriate pay scales and promotion, was introduced along with the title 'navigator'.

On board *Warrior* the master would have had an assistant master, a master's assistant and a number of midshipmen training as navigators. *Warrior*'s master on commissioning was the 38-year-old George Blakey, a very experienced and innovative individual who invented an ingenious distance-measuring device while with the ship.

Blakey served in *Warrior* throughout her first commission and he proved a competent navigator. In his later career he was appointed one of the very first staff commanders, finishing his career as the Queen's Harbourmaster at Milford Haven.

Officers

Cochrane and Tryon spent their initial time in the harbour flagship *Fisgard* at Woolwich as *Warrior* was completed by the Thames Iron Works. Cochrane was permitted to select his officers, but he could not select warrant officers as these were appointed by the Admiralty. The most senior officer was Henry Phillimore who had just spent 3 years serving on the Pacific Station in the flagship *Ganges*. He was an experienced officer who had specialised in gunnery and was appointed first lieutenant; the second lieutenant was Joseph Wilson who had just finished serving in the Mediterranean flagship *Marlborough*; returning from the Pacific Station was George Parker who was a qualified watch keeper and became third lieutenant. The fourth and fifth lieutenants were Henry Perceval and Noel Digby.

These officers were known as the Executive or Military Branch. Other appointments were known as Civil Branch officers and these

comprised the surgeon Samuel Wells, assistant surgeon William Asslin, the chief engineer William Buchan who had returned from the Pacific with Henry Phillimore, his assistant engineer William Glasspool, the paymaster John Nicholas de Vries, and the chaplain and naval instructor the Revd Robert Jackson.

Sub-lieutenants and midshipmen

Midshipmen achieved their place in the navy by patronage either from a senior officer, who may have been a relation, or their parents knowing a senior officer. The British class system was adamant that these young men came from the right social background: the aristocracy, the upper middle class, the Church or the sons of military officers. Those from humble backgrounds and most definitely those from 'trade' had no possibility of being accepted. Midshipmen could be in the navy as young as 13, firstly in a training ship, and at 15 years went to sea either as a cadet or midshipman and then could serve up to 6 years in a ship, specialising in gunnery, navigation and general seamanship. Midshipmen came under the discipline of the ship's chaplain who shared the gunroom with them along with the assistant paymasters and clerks. Each midshipman would have a chest that contained his personal belongings and the tools of his adopted specialism. They slept in hammocks slung above their chests in the chest room. Every midshipman had to go aloft daily and keep a journal of their activities and training.

For centuries, water – the most fundamental source for sustaining life – had been impossible to store on a ship for any length of time without it becoming contaminated. Therefore alcohol became the alternative. During the Napoleonic Wars, ships of the Royal Navy issued up to 8 pints of beer to each member of the crew every day. This may well have been weak beer, but there was little alternative, especially for any ship at sea for any length of time. The problem was, however, that beer could also spoil. When beer ran out, it could be substituted by a pint of wine or half a pint of spirits, depending on what was locally available.

When Admiral Edward Vernon took command of the Caribbean Squadron in the 1740s he discovered on his arrival that traders in Jamaica and Trinidad were supplying large quantities of rum to ships that were under his command. Half a pint a day was being issued per man and in some cases individuals were consuming up to a pint as an illicit trade was rife. A story from the era says that sailors would prove the strength of the rum by dousing gunpowder with it and seeing if it would still burn, thus verifying that rum was at least 55% proof.

Vernon quickly became aware that, after consuming the spirit, many of his crews were incapable of carrying out their duties and therefore ordered the dilution of the rum in the proportion of half a pint to one quart of water (1:4). His order stated:

Whereas the pernicious custom of the seamen drinking their allowance of rum in drams, and often at once, is attended by many fatal effects to their morals as well as their health, the daily issue of half a pint is to be diluted with a quart of water to be mixed in one scuttlebutt in the presence of the lieutenant of the Watch.

Vernon always wore a cloak made from grogram (grosgrain), a heavy ribbed fabric made from silk and mohair, and as a consequence earned the nickname of 'Old Grog'. Very quickly the diluted rum became known as 'grog'. The rum ration was also split into two controlled issues, one in the morning and one in the early evening. In 1756 lemon or lime juice was added to prevent scurvy.

The ration was cut in half in 1823 and again in 1850 to half a gill, or one eighth of a pint watered down with three parts water. At this time 'grog money' was introduced whereby teetotallers would be paid 1s 7d per month additional pay for their abstinence.

By the time of *Warrior*, drinking water was available and another beverage had made its mark – a cup of tea. The rum ration was reduced yet again in 1937, though warrant officers, chief petty officers and petty officers could draw their tot neat. The issuing of rum was discontinued in the Royal Navy in July 1970.

RIGHT Rum.

They were forbidden all alcohol, including the rum tot, until they reached 18. Midshipmen would be tutored by the officers and work with the warrant officers and senior members of the crew where they would be given responsible tasks to undertake.

One midshipman's ability proved invaluable over a hundred years later when *Warrior* was taken to Hartlepool to begin her restoration. Of course the restorers had information about her construction and of the engines and boilers, but the minutiae of life on board were missing. What sort of furniture was in the wardroom, the carpenter's workshop, cabins, the mess tables and the buckets and pots and pans, and where were they located in the ship?

But a young midshipman, Henry Murray, had been given a project in 1861 to draw the ship. He produced the most detailed plans of all the decks showing every particular that proved invaluable in the restoration of *Warrior*. An enlarged copy of his plan is displayed in the gunroom where he probably drew it all those years ago.

Warrant officers, chief petty officers and petty officers

'A captain of a ship is second only unto God. But the warrant officer is God.' Though said in jest, there is an underlying truth that warrant officers, chief petty officers and petty officers had achieved their position through long experience and ability, and though they demonstrated deference and respect to officers it would be a very foolish and naïve officer who did not heed the advice of a senior warrant officer. These men knew more about ships and how they worked than the officers collectively and indeed the rank of warrant officer dates back to the very early Tudor days where captains of ships were noblemen appointed by the Crown and who had little or no knowledge of what they were doing. Experienced sailors were therefore appointed by 'warrant'.

The warrant officers, chief petty officers and petty officers occupied the cable deck forward and had their own cook and the luxury of cubicle heads (toilets).

Seamen

Of the 430 crew, fitness and physical ability was the predominant requirement. The three major tasks were going aloft to manage the sails, manning guns and stoking the boilers.

Orderliness was the very core of the navy. Captains, crews and ships were judged and rated on their speed at sailing and gunnery and overall smartness. This was taken to extremes by some senior admirals. Training and routine was continuous. Uniforms were now issued to sailors and an individual's appearance reflected on the whole ship.

LEFT The 'ditty box', where a sailor kept his personal possessions. *(Author)*

Uniform and kit allowance
sennet hat
soft hat
2 × blue suits
2 × white suits
comforter
2 × towels
4 × collars
2 × white fronts
2 × pairs of socks
2 × handkerchiefs
2 × silks
jack knife
blanket
2 × lanyards
shoe brushes
clothes brush
ditty box
name stamp
kitbag
2 × hammocks

BELOW A sailor's clothing issue. *(Author)*

The 'ditty box' was a small wooden box or bag that each man was issued with. In it he could keep personal effects. The word 'ditty' was a corruption of the Hindi word 'dutty', which means calico cloth. By the time *Warrior* was in service, photography had been invented and was becoming popular, and many boxes contained pictures of family and loved ones.

Maintaining discipline was a continuous task. Though shore leave was a privilege that most of the crew honoured by returning on time, some instances of being 'absent without leave' were inevitable and the ship's book would be marked 'R (run) a deserter', or 'RQ (run query) possible deserter'. Both marks meant a stoppage of pay and privileges. Arthur Cochrane wrote a number of letters to the Admiralty in regard of genuine absentees caused by sickness and unavoidable circumstances seeking leniency.

Those of the crew who did not receive leniency were the ones who came back on board drunk and incapable of their duties, or those who sold their issued clothing when ashore, which appeared to have been a common practice.

RIGHT Officers and crew of HMS *Wasp* around 1850. The unusual choice of hats is of interest: the more unorthodox ones are probably worn by the Civil Branch officers, such as engineers or the surgeon, who at that time would not have worn naval uniform. *(NMRN)*

Engineers

Engineers were the new breed of men who were increasingly found on board Her Majesty's ships. Until *Warrior*'s heyday, steam engines on warships were little more than supplementary to sail and were used for entering and leaving harbour as well as tactical manoeuvring when in action. But *Warrior* was different in that she could continuously transit under steam and therefore carried sufficient coals for that duty.

There were three trades within the Engineering Branch: engineers in the purest sense (who ran and maintained the engines); the stokers (who managed the boilers, keeping them at optimum performance); and coal trimmers (who brought the coal from the bunkers to the front of the boilers).

Royal Marines

The Corps of Royal Marines are the elite of the British armed forces. They have a history that can be traced back to the formation of the English Army's Duke of York and Albany's Maritime Regiment of Foot in October 1664.

Known within the navy as the 'Royals' or 'soldiers', their training was, and indeed remains, robust. In the mid-19th century it was noted that the Royals attracted a high proportion of agricultural workers, who were tough, solid and fit individuals. Discipline, coupled with steady performance, was the reputation they brought aboard a ship.

The detachment assigned to *Warrior* was drawn from the Royal Marine Artillery Division that had only just been formed in 1859. It was the first time such a detachment had been assigned to a ship and it was acknowledged as an experimental posting. They occupied the aft mess tables and manned the adjacent 110lb Armstrong guns and 68lb cannon. The Royals' location on the gun deck was not coincidental but followed another naval tradition. In 1797 there had been a mutiny at the Nore and Spithead over the appalling living conditions on warships. It was brutally suppressed, but it succeeded in bringing about changes. From then on the Royal Marines' mess was always positioned

A TYPICAL DAY IN THE ROYAL NAVY, 1860	
3.30	Daily routine
4.00	Coil ropes
5.45	Scrub decks
6.00	Re-set sails
6.30	Stow hammocks
7.15	Breakfast
9.00	Cleaning
9.30	Prayers
11.30	Drills
12.00	Up spirits' rum issue
13.00	Dinner
13.30	Roll call
16.15	Drills
17.00	Supper
19.30	Quarters
20.30	Stand by hammocks
Rounds conducted throughout the night	

between the crew and the officers as Marines were looked upon as ships' policemen, maintaining order and discipline.

The Royal Marines Artillery uniform was a blue jacket, white trousers and pillbox hat with the Royal Marines badge. 'Royals' were not permitted to grow beards as many in the crew did, but older Marines did have moustaches or significant side whiskers.

The marine captain on *Warrior* was Henry Mawbey, his first lieutenants were Herbert Everitt and Hastings Owen, and in addition he had 3 sergeants, 3 corporals and bombardiers, 2 drummers and buglers and 114 gunners.

They came with an enviable reputation; in training it was reported they had achieved five rounds in 2 minutes from an Armstrong gun hitting the target every time. With this rate of fire it was almost certainly from the 40lb Armstrong gun and not the 110lb.

By tradition the 'Royals' were gangway sentries when in harbour and one would be the captain's sentry posted outside his cabin. In reality he was the captain's 'runner', delivering messages around the ship.

But the 'experiment' of an entire artillery contingent was short-lived and in early 1863 Lieutenant Owen and 41 of his artillerymen were withdrawn and replaced with Royal Marine

Infantry, *Warrior* having two-thirds artillerymen and one-third infantry. When these new 'Royals' arrived, the ship's blacksmith, Watling, lifted each man's leg and wrenched the metal heel off his boots. The holystoned and perfect decks were not going to be ruined by heavy-footed Marines!

Food, stores and pay

Though *Warrior* did not undertake long voyages in her active life, she was supplied with sufficient stores to enable her to be at sea for three months. She carried provisions, water, ammunition, powder, timber, spares and all the materials for the crew to live and to be self-sufficient and to make minor repairs from any action or storm damage the ship might encounter.

Stores are known in the navy as 'victuals' and 'victualling', the process of embarking the food, clothing, bedding and all the paraphernalia that a ship would need to function, feed and support the vast crew.

Warrior's provision stores are aft on the orlop deck along with a bread store and spirit room. Other stores such as the engineers', boatswain's, sail room, carpenter's and an issuing room are spread around the ship.

Cattle pens were erected on the upper deck or on the cable deck for the stowage of live cattle and poultry. There are even stories that one of the cutters was used as a chicken run. Vegetables were stored in canvas cages, again on the upper deck.

Within the provision stores there would be 19 tons of salt beef in casks; 10 tons of flour, butter, suet, soap and tobacco. There were 10 tons of sugar, 6 tons of peas, 2 tons of vinegar and lemon juice and 4 tons of rum in casks. Amidships were stowed 8 tons of candles, raisins, tea, chocolate, tins of preserved meat and cheese.

Spare clothing, known in the navy as slop clothing, was carried in the slop store opposite the issuing room on the lower deck. Various sizes of uniform were stocked to cater for all aboard. The Marines had their own slop store.

Candles were used in huge quantities. Apart from some oil lamps, candles provided the only illumination on the ship. They came in an assortment of sizes with various duties attached: 'police', 'fighting', 'signal', 'eights' and 'twenty-fours'. The largest candles were the 'police' type that burned for more than 12 hours and illuminated important spaces around the ship. 'Twenty-fours' burned for only two hours, while 'eights' were eight candles to the pound and lasted three times as long as a twenty-four. Fighting candles were inserted into lanterns between each of the guns and signal candles, as their name suggests, were used in signal lamps. Lamps and candles were stowed in the steering flat and were managed by the lamp trimmer.

Victuals

There is a world of difference between the horrific stories about the quality of food that have emerged from navy ships during the Napoleonic Wars and the food that those on board *Warrior* and in the wider Victorian navy enjoyed. Hygiene, the health of the crew and the need to sustain their physical wellbeing was now understood and so adequate quantities of better food was seen as key to an efficient navy.

Warrior was probably luckier than other ships in that she did not endure long deployments overseas. She remained in British waters nearly all her active life with only occasional voyages to the near continent. Provisioning of both water and food was never far away.

The daily and weekly allowance of food per man is given in the table opposite. The captain and officers received the same allowance but could supplement it by purchasing additional provisions. This practice was extended to the crew and each mess could club together to buy additional food. This was, however, discretionary based on the amount of victualling there was left before the next port of call. The ship's steward was in charge of the daily issue of provisions.

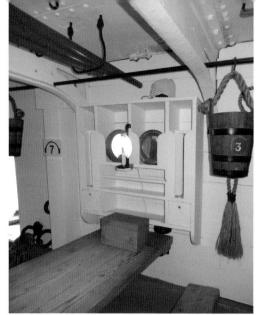

Daily per man	Foodstuff	Notes
1½lb	Biscuits or soft bread	
⅛ pint (gill)	Rum	3 parts water:1 part rum
2oz	Sugar	
1oz	Chocolate	
¼oz	Tea	
1lb	Fresh meat	When available
1lb	Fresh vegetables	When available
1lb	Salt pork or salt beef	In lieu of fresh meat
⅓ pint	Split peas	
9oz	Flour	
¾oz	Suet	
1½oz	Currants or raisins	

Weekly		
¼ pint	Oatmeal	
¼ pint	Vinegar	
½oz	Mustard	
¼oz	Pepper	

The 34 mess tables, arranged on the length of the gun deck, accommodated 18 men each. Naval terminology again prevails as all the equipment that was assigned to each mess had a specific name and duty. Generally known as 'mess traps' or 'mess gear', it comprised two mess kettles, two meat dishes, two hook pots, one pepper dredge, one salt jar, two soup ladles and one metal drinking cup between four men. All were marked with the mess number.

Each man had his own metal plate and metal basin. Issued as part of his personal kit was a clasp knife that he used both for work and for cutting up his food. Fingers were still used between plate and mouth. A spoon was the only cutlery issued and this was for eating soup or porridge.

All the plates, basins and spoons were kept in a rack at the end of the mess table on the ship's side. There was a tablecloth made from canvas and vinegar was used as the cleaning agent.

'Cooperage' is the collective word for buckets and basins made from wood. Each mess had a wooden bread barge, wash buckets, fire buckets, vinegar barricoe, spitkid and latrine bucket.

Warrior still retained a traditional and practical naval routine for catering for the 700 men aboard. Of the 18 men on a mess table, one would be nominated as the mess cook. If the mess was democratic the mess cook would change each week, but if one man was found to be proficient, he may well have the job more often than not. He would go to the issuing room on the lower deck and draw the food ration for

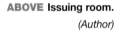
ABOVE Issuing room.
(Author)

ABOVE RIGHT Boiling
bag for prepared
vegetables or meat.
(Author)

RIGHT Two of the
few artefacts to have
survived are the
original galley stove
and the brick floor.
(Author)

his mess each day, then bring the food back to his mess table and prepare it. Anything requiring baking would go into the mess kettle and anything that needed to be boiled went into a net bag – each with the mess number marked on it – and then taken to the galley. This was an efficient way of feeding the large crew. The captain, commander, officers, midshipmen and senior non-commissioned officers all had their own cooks. This practice of a mess cook preparing the food was still a routine in the Royal Navy as late as the 1960s.

Breakfast was usually 'kye' (cocoa), biscuit or bread and preserves or cheese. Dinner was soup or porridge, boiled meat and vegetables, suet pudding and tea or kye. The evening meal was usually a cold version of the midday meal.

The galley did little more than provide heat and hot water in the large copper pots on top of the stove. The coal-fired stove was continuously lit, the fire only being extinguished if gunpowder was being embarked or the ship was in action. The galley had to be ready by 4.00am for breakfast at either 6.00 or 7.00am. Dinner, the main meal of the day, was at 1.00pm and supper was at 5.00pm.

Pay

The paymaster was the ship's accountant and his office was between the captain's and commander's cabins on the half deck. He

LEFT The galley stove still works and is occasionally lit. (Author)

controlled money, provisions and clothing. He had a staff that comprised four assistant paymasters and at times two further paymasters under training.

Pay was distributed to the officers and crew monthly in cash and this was a formal activity. Off-watch seamen reported to the pay office and, at the command, a seaman took off his hat so that his wages could be put into it.

The table below shows the annual pay for naval personnel of all ranks for 1861:

Rank	£ per annum
Rear admiral	1,095
Captain in command	584
Captain	365
Commander	301
Lieutenant	182
Master	273
Engineer	164
Boatswain	120
Midshipman	32
Master at Arms	41
Ship's cook	36
Gunner's mate	36
Stoker	36
Coxswain	33
Able seaman	29
Ordinary seaman	23
Boy second class	9

A Bank of England guide suggests a 110% inflation factor is applied for a 2017 equivalency.

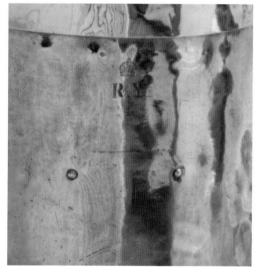

ABOVE LEFT **Butchers' tools.** (*Author*)

ABOVE Galley pans. (*Author*)

LEFT Many of the pans were donated to *Warrior* from the royal yacht *Britannia* and carry the royal stamp 'RY' surmounted by a crown. (*Author*)

LEFT The galley chimney passes up on to the upper deck and still functions when the stove is lit. (*Author*)

Health

In Britain in the early 19th century there was little access to medical assistance, either medication or clinical, for the general population. Even those from the privileged classes would be subjected to many medical practices that would be highly questionable to modern eyes.

The nation was in transition from an agricultural-based economy to an industrial powerhouse and there was subsequent growth of the urban population where water and food quality was suspect and sanitary provision comprised little better than cesspits. There was not much understanding of the causes of disease and sickness and how it spread.

Between 1849 and 1853 over 25,000 Londoners had died from cholera alone. Typhus and typhoid were rampant and the prevailing view of medical 'experts' of the time suggested that all these illnesses were caused by 'foul air'. Even when Dr John Snow had proved that contaminated water was the cause, he was not

believed. But 1858 became known as the year of the 'Great Stink' where the River Thames had become nothing more than an open sewer.

With these widespread health issues in the cities and towns and the shambles of the Crimean War, where more troops died from disease than military action, changes had to be made. The publicity that Florence Nightingale's enlightened activities received with the horrors that she had faced in the Crimea achieved much in galvanising the press and public opinion, urging the government to do something to improve the health not just for the military but for the population as a whole. Health and welfare had started to become a political issue.

The Chancellor of the Exchequer, Benjamin Disraeli, had proved difficult over the projected naval costs and had forced the Admiralty's hand in building only *Warrior* rather than the numerous ships first requested to match France's naval expansion. But the money for *Warrior* was infinitesimal compared to the sums that were required for what has been described as the greatest engineering feat the world had seen: the building in 1858 of London's sewers under the guidance of Joseph Bazalgette. The nation was at last becoming health conscious.

In the navy little attention had been paid to the wellbeing of crews. As long as they did the task, they met the need. If they became ill, got injured or killed, you got another man to replace them. Medical provision was basic.

The immediate impact from the Crimean War for the navy was to raise the surgeon to the status of an officer. In 1861 Queen's Regulations and Admiralty Instructions were issued that introduced hygiene and cleanliness standards throughout the fleet. Carbolic soap was used and chloroform was now available to sedate patients on the operating table.

Medical staff

Warrior had a surgeon, two assistant surgeons, a sick berth steward and two attendants. Men on 'light duties' could also be assigned for cleaning and bringing meals to the patients.

The surgeon would have attended an approved medical school to acquire academic and practical qualifications before he could join the navy. He would have had to present certification for the following:

Anatomy	General anatomy	18 months
Surgery	General surgery and military surgery	18 months
Theory of medicine	Attendance of lectures on the theory and practice of medicine	up to 18 months
Clinical lectures at a hospital	On the practice of medicine On the practice of surgery	6 months 6 months
Chemistry	Lectures on chemistry Practical chemistry	6 months 6 months
Materia Medica	Therapeutic properties	6 months
Midwifery	Presentation of certificates stating number of cases personally attended	6 months
Botany		3 months

One of the major drawbacks surgeons had when on a ship was their isolation from sharing their knowledge with other surgeons and therefore a wide skills and capability gap appeared among surgeons in the fleet. Those on 2- and 3-year deployments in a ship in the far reaches of the British Empire probably had little or no updates on medical advances or treatments that were occurring.

Routine

Every man joining the ship was required to undergo a medical examination and records were kept of each individual: his height, weight and any medical conditions he may already have. Vaccination against smallpox was obligatory as required by the 1853 Vaccination Act, but over a quarter of the seamen joining the ship in 1861 had suffered from the disease.

Each morning the surgeon presented to the captain the 'sick list' and on *Warrior* this ranged from 25 to 90 individuals, the majority of which

ABOVE Sickbay cots.
(Author)

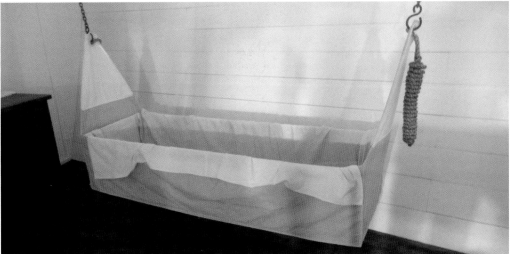

LEFT If high numbers of casualties were experienced, temporary cots could be rigged up in the sickbay. *(Author)*

RIGHT The surgeon's overall was more for keeping his uniform clean when operating. He would not change overalls between patients. The bandages have been used, then washed and are hanging up to dry ready for the next patient. *(Author)*

BELOW The surgeon's instruments and tools. *(Author)*

would be 'walking' sick who were diagnosed, treated and put back on to their mess.

The routine followed this procedure:

7.00am	Sickbay inspection by the surgeon
7.30am	Dispensaries open for medicines
9.00am	Report to captain with sick list
9.30am	Examination of new cases and minor operations
4–6.00pm	Dispensary open for medicines, usually attended by one of the assistant surgeons

As a warship, *Warrior* conducted exercises for battle and the inevitable high casualty numbers that could be expected. The wardroom was converted into what was known as the second first-aid post, the sickbay being the first. Each first-aid post was run by one of the assistant surgeons with the surgeon himself moving between them as required. The paymaster, naval instructor, chaplain and any staff not assigned other duties would assist. A temporary mortuary would be established.

SURGICAL INSTRUMENTS, 1862

3 × amputating knives	pair of probe scissors
1 × amputating saw	curved bistoury with button (narrow surgical knife)
1 × metacarpal saw	long probe
2 × catlins	pair of bullet forceps
2 dozen curved needles	2 × prolongs (instrument for pushing obstructions down oesophagus)
2 × tenaculums (surgical hooks)	½lb of surgical thread
6 × petit screw tourniquets	1 × paper of needles
pair of bone nippers and turn screws	case with lift-out apparatus for restoring suspended animation
3 × trephines (saw for the head)	pocket instrument set
pair of forceps	6 × lancets (used for opening veins)
elevator	2 dozen boughies
2 × trocars	2 pint pewter clyster syringes (clyster is a liquid injected into the intestines)
2 × silver catheters	6 × small pewter syringes
2 × gun elastic catheters	2 × sets of splints
6 × scalpels	12 × flannel/linen rollers
small razor	2 × 18 tailed bandages
key tooth instrument	20yd of tourniquet web
gum lancet	60yd of tape
2 pairs of tooth forceps	cupping apparatus of scarification and 6 glasses
punch	Fahrenheit thermometer
2 × seton needles	

The surgeon would have to buy his own instruments against an Admiralty requirement list and any other equipment that he believed he needed.

Despite the improvements in medical facilities, deaths were inevitable and in 1862 the following are recorded in *Warrior*'s medical returns by Surgeon Wells:

1 × inflammation of the lungs and pleura;
2 × disease of the stomach and bowels;
1 × disease of bladder and kidneys;
2 × wounds and injuries;
6 × drowned.

The last was a tragedy when a boat bringing crewmen back to the ship capsized. Many sailors of this period could not swim.

Chapter Six

A new beginning

'If every warship in the 19th century still existed and was available for preservation, Warrior would still be my first choice.'

Sir John Smith
Vice-president, The Maritime Trust, and founder of the Warrior Preservation Trust.

OPPOSITE HMS *Warrior* **before restoration.** *(WPT)*

Throughout the summer of 1883 the fate of the decommissioned *Warrior* hung in the balance, but behind the scenes an influential ally was at work in the guise of Captain George Tryon, Secretary to the Board of Admiralty. He advocated that *Warrior* and her contemporaries still had enormous capability as they were proven seaworthy craft, strong, durable and a hull form capable of high speeds. For *Warrior*, proposals were submitted to install new compound machinery, convert her to twin screw to achieve speed of over 15kts, upgrade the armour to bow and stern and install 40 guns. All for less than £130,000.

Why was Tryon such an advocate for *Warrior*? Possibly because, as we have already discovered, he had been her commander under Cochrane during her first commission. Everyone from the First Sea Lord to the Director of Ordnance agreed. A further survey of *Warrior* was undertaken and confirmation was received that she was in remarkably good condition.

Yet nothing happened. *Warrior* was moored in Fareham Creek a short distance from Portsmouth Harbour with four watch keepers on board. Theoretically, she was in reserve to be brought into service in the event of war, though by 1887 it was generally accepted that old iron ships would be only useful as transports in any future conflict.

RIGHT John 'Jackie' Fisher. First Lord of the Admiralty, 1903. *(NMRN)*

Despite her dormant state, *Warrior* kept appearing in the Admiralty list of ships of the Royal Navy and she also kept changing her designation. Firstly as a 'screw battle ship' with a distinct gap between the word 'battle' and 'ship', then in 1892 she was assigned as a 'first-class armoured screw cruiser'. This was possibly in anticipation of her proposed conversion to a cruiser. By this time *Warrior* was laying uncared for and forlorn. Her wooden masts had been found to have rot and had been removed. She was all but derelict.

In 1899 the Admiralty received some criticism in the press when they published that year's strength of the fleet, including *Warrior* and other laid-up ships as armoured cruisers. Accusations of paper accounting were made. Her old advocate George Tryon tragically lost his life in 1893 when in command of the Mediterranean Fleet. His flagship *Victoria* collided with the *Camperdown*, the *Victoria* sinking and taking Tryon with her.

But there was another saviour in John Arbuthnot Fisher, known as 'Jackie' Fisher, who had been the gunnery officer on *Warrior* and was now the Commander-in-Chief Mediterranean Fleet. Fisher was a mesmeric character, a superb administrator and innovative military thinker. From the Mediterranean Fleet, Fisher was promoted to become Second Sea Lord in 1902.

Though Fisher had always been a gunnery exponent he recognised that the torpedo, when mounted in small fast boats, was the new threat and an innovative weapon. The threat was answered by the building of torpedo boat destroyers. Companies such as John I. Thornycroft & Co. in Southampton and Yarrows in Glasgow produced considerable numbers of original groundbreaking designs.

These ships, of around 400–500 tons, could reach 30kts. Formed into flotillas they were located strategically around the country. The Portsmouth flotilla needed a headquarters and depot to look after them. Fisher saw the answer in *Warrior*.

Warrior's 'conversion' was the removal of her engines and boilers and the installation of workshops, stores, fuel and torpedo storage.

She was commissioned in July 1902 in Portsmouth with 7 officers and 100 crew. She had 8 destroyers alongside.

Successful though this was, it did not last long as another task was emerging in the new age of technology. A torpedo school was being established and *Warrior* was chosen to be part of it. After 2 years' service as the destroyer depot ship, she was taken to Southampton for conversion to a floating workshop for the new HMS *Vernon* torpedo school.

Warrior had her main deck roofed-over to provide classrooms, offices and other facilities including a chapel. Boilers and electrical generators were fitted. A doorway was cut through the aft transverse armoured bulkhead, giving better access through the lower deck. The original boiler room became a machine shop. Returning to Portsmouth, she was moored between the old three-deck wooden warships *Donegal* and *Marlborough*,

each connected by a gangway and renamed as *Vernon I*, *Vernon II* and (*Warrior* losing her name) *Vernon III*. She became the generator ship of the HMS *Vernon* torpedo school.

In 1903 two new armoured cruisers were ordered and inherited the names *Warrior* and *Black Prince*. By this time Jackie Fisher was First Sea Lord and through his instigation and drive HMS *Dreadnought* entered service in 1905, having taken just a year to build. She was the first big-gun battleship that brought about one of the most radical transformations in naval warfare, just as *Warrior* had done when she was constructed some 45 years earlier. *Dreadnought* was built in Portsmouth Dockyard on a slipway almost opposite to where *Warrior*, as *Vernon III*, was moored. Two great naval innovations from different eras sat facing each other.

ABOVE **HMS** *Dreadnought* passing **HMS** *Victory*. *(NMRN)*

The 'Dreadnought' became the datum for all navies of having 'Dreadnoughts' or obsolete 'Pre dreadnoughts' in their fleets. Fisher's active mind must surely have recognised the association between his old ship *Warrior* and the *Dreadnought* he had created.

Wireless telegraphy was yet another new technology that the *Vernon* school undertook. Very much a coming facet of the modern navy, *Warrior* had Marconi's wireless equipment installed in 1908.

The world was changing and war was approaching. Germany's aspirations and her military capacity had grown exponentially. Her navy was becoming a direct challenge to Britain's historic and divine right to 'rule the waves'.

When the First World War began in 1914,

the old *Warrior* did her duty, passive though it was, seeing many sailors pass through her decks as they trained in the professions needed for the greatest of conflicts. But there was no excitement, no battles and no glories for her. The ultimate sea battle of the First World War was the Battle of Jutland in May 1916; in it, the *Warrior* and *Black Prince* successors both paid the ultimate price and were sunk. Light cruisers were no match for German heavy cruisers. Jutland was a victory for Britain, despite the heavy losses, as the German Navy ran for home never to put to sea again until their surrender in 1918.

When the 'war to end all wars' was over, there was an inevitable reduction in the navy. The *Vernon* school had a new home ashore at Gun Wharf in Portsmouth (this is now the Gun Wharf Quay shopping centre, adjacent to the restored *Warrior*).

As a result, the old wooden three-deckers attached to *Warrior* were sold for breaking up. For a short time *Warrior* was used as a floating annexe to the shore-based *Vernon* and then her seemingly ultimate fate was announced.

On 31 August 1924 *Warrior* was decommissioned and offered for sale for scrap. She was towed to Fareham Creek where she lay for two years awaiting her end. During that time, *Warrior*'s sister ship *Black Prince* was sold and scrapped at Dover. Many ships were being scrapped, which led to the price of recycled metals plummeting and therefore *Warrior* was of little monetary value.

RIGHT The light cruiser *Warrior* launched in 1905 and was lost at the Battle of Jutland in 1916. *(NMRN)*

H.M.S. WARRIOR.
Armoured Cruiser, 13,550 tons.
Cost £1,200,000; Length, 480 feet; Beam, 73½ feet; Draught, 27½ feet; Speed, 22½ knots. Armed with Six 9.2 in.; Four 7.5 in.; Twenty-nine Small Quick Firing Guns, and Three Torpedo Tubes.

Once again her hull was inspected and found to be dry and in excellent condition. There were desultory talks of preserving her, but post-war feelings were not conducive to the idea and nothing came of it. As a result of the survey, the Admiralty found a task for her. She was redesignated as an 'oil fuel pontoon hulk' in 1927.

In March 1929 she was towed from Portsmouth to Milford Haven in South Wales and into the River Cleddau to become the fuelling pontoon of the Llanion Admiralty oil terminal. It would be *Warrior*'s home for the next 50 years.

Llanion oil depot

Prior to *Warrior*'s move, she had been stripped of all her equipment, including the roofed-over main deck, and it was a very empty ship that arrived in South Wales. The Llanion oil depot was adjacent to the river, and from there two pontoons and gangways extended; *Warrior* was moored on to these. The fuel pipes ran through a long tunnel from the depot over the pontoons and on to *Warrior*. The tidal rise and fall was considerable, up to 24ft (7.3m) with a strong current, but there was a good depth of water of up to 48ft (14.5m) on the tide. About a dozen ships a month were refuelled.

The poop deck accommodation that had been added when she was in the Reserve Squadron had been retained and this was converted into accommodation for the ship keeper and his family. Though water had been connected, there was no electricity and lighting was by oil lamps. Heating and cooking was by coal. Not much had changed in that respect from when *Warrior* was new.

Warrior was periodically docked and though the hull was still in good condition, deterioration of the upper deck resulted in a thick layer of concrete being laid.

The Second World War came and with it an intensity of activity. She took on a dual role, still as the fuelling pontoon, but also a base for coastal patrol craft. A wooden Nissen hut was built on her upper deck as their headquarters.

The Luftwaffe attacked Pembroke Dock on 19 August 1940 when three Junkers Ju88 bombers targeted the oil terminal. The attack sparked what has been described as Britain's largest blaze since the Great Fire of London,

ABOVE Llanion oil depot, Pembroke Dock. *(WPT)*

BELOW Llanion oil depot, Pembroke Dock. *(WPT)*

BELOW The oil depot burning after the attack by the German Luftwaffe on 19 August 1940.

RIGHT **The Waitimus family. Jennifer (second right) was born on board *Warrior* on 27 May 1950.** *(WPT)*

with the oil tanks burning for 18 days. The damage to the oil tanks was so significant that the fuel was contained behind large earthen walls around the devastated site. *Warrior* escaped unscathed.

In 1942 she had the indignity of being renamed 'Oil Fuel Hulk C77' – not the most glamorous of names – but necessary because

the 'Warrior' name was needed for a new aircraft carrier being built in Belfast for the Canadian Navy. At the end of the war and with peace returning to Pembroke, Oil Hulk C77 continued to provide her fuelling services.

In May 1950 the ship keeper's wife, Mrs Waitimus, provided *Warrior* with an historic moment when she gave birth on board to a daughter, Jennifer.

Warrior was still regularly docked either at Pembroke or Devonport. The surveyor's report stated: 'After close examination of the plating, rivets and frames, the very good condition of the hull structure, coupled with the excellent standard of construction, left no doubt that *Warrior* would outlive a lot of warships built today.'

In our current age we are probably more aware of our heritage and the preservation of our history than we have been at any other time, and so it is difficult to look back and understand that, as late as the 1960s, ships of over 100 years or more old were regularly sent to the scrapyard. HMS *Agincourt*, built in 1865, had languished as a coal hulk until 1960 when she was sold for scrap. This left only *Warrior* now a century old.

But with *Agincourt* going to scrap, people were slowly becoming aware that *Warrior* was the last of the ships from that great era of transformation that had changed Britain and the world. She epitomised the nation's capabilities, her technological advances, the social changes she brought and the absolute confidence that had existed in building her.

The problem was that *Warrior* had no glorious history, no battles, no associated heroes, no great pedigree. As a consequence, she was only known to those who took an interest in naval history and who realised how important she was. To the majority of the general public she was an unknown.

On 4 August 1960 the battleship HMS *Vanguard*, built in the Second World War, was towed from Portsmouth Harbour to the breaker's yard. The press and public believed her to be the last Royal Navy battleship. Except she was not the last – *Warrior* was. As the author John Winton said: '*Warrior*, the first and last battleship.'

During 1967 Frank Carr, who had been the director of the National Maritime Museum at

BELOW ***Warrior* in dry dock at Pembroke Dock.** *(WPT)*

Greenwich and HRH the Duke of Edinburgh suggested that *Warrior* should be brought to London and become part of a proposed yacht harbour that was to be built at Thamesmead, accepting that considerable work would be necessary to restore her. The scheme would have included community facilities on board as well as a restaurant. There was the view that *Warrior* was a London ship, having been built on the Thames.

The Ministry of Defence permitted *Warrior* to be surveyed by Lloyd's Register and representatives from the National Maritime Museum. Her hull was found to be in excellent condition, but again the extensive work that would be needed to restore her was noted.

An early meeting to discuss *Warrior*'s future was held at Buckingham Palace in 1968 with the Duke of Edinburgh in the chair. Attending was the historian Sir Arthur Bryant, Sir Michael Cary from the Ministry of Defence, John Smith who was the Conservative MP for the cities of London and Westminster and Frank Carr.

Emerging from this meeting was a proposal for a National Maritime Trust that would preserve ships in the same way that the National Trust did for historic buildings. Prince Philip asked John Smith to form the Trust and this he did together with a promise of providing a financial contribution to start things off.

In August 1968 Prince Philip visited *Warrior* at Milford Haven. This brought welcome publicity for the ship and her preservation and all discussions were now positive. There was no longer any talk of disposal.

The Maritime Trust was formally inaugurated in October 1969 with Prince Philip as president, John Smith as vice-president and Admiral Sir Patrick Bayly as director. John Smith was very much the driving force. A banker prior to his election to parliament, he had founded the Manifold Trust in 1962 that raised money for charities by buying long leases which were close to their dates of expiry. Though this was a rather speculative enterprise, it proved very successful, producing funds for many of his charitable interests. He was involved in canal restoration and was an outstanding influence behind the preservation of HMS *Belfast*, SS *Great Britain* and ultimately *Warrior*.

Significant issues had to be recognised.

Firstly *Warrior* was still committed to her task as a refuelling jetty at the oil terminal and secondly the Thamesmead Marina Project that would accommodate *Warrior* had become increasingly doubtful, with estimates for her berth alone already exceeding £1million. As a result, the scheme quietly disappeared.

The London Borough of Newham stepped in with suggestions of bringing *Warrior* to the Royal Victoria Docks with the association of her having been built and fitted out there. It was a good idea, but like Thamesmead, was not financially viable for a local authority to fund.

At around this time Portsmouth showed interest in offering *Warrior* a home. As to where *Warrior* would eventually be housed was academic as the real issues to be confronted were finding money, acquiring her from the MoD and finding both a location and the people to undertake what would undoubtedly be the enormous task of restoring and rebuilding her.

The arrival in Bristol of the SS *Great Britain* in June 1970 from her long stay in the Falkland Islands gave the preservation of historic ships a real boost in attracting media and public interest in the country. The restoration of the *Great Britain* served as the learning curve for *Warrior*.

In July 1974 *Warrior* was docked at Milford Haven for what was to be her last docking while owned by the MoD. Yet again she was fully surveyed. It was noted that water had to be pumped out of her on a fairly regular basis. On examination this was found to be rainwater

intent would be for *Warrior* and Sir Patrick Bayly established the significant starting point by reaching an agreement with them that on the closure of the oil depot *Warrior* would be offered to the Maritime Trust as a gift. However, the Trust had to provide to the government evidence that they had the money and ability to restore the ship. She was not to lay forlorn for lack of funding.

It was with irony that, having refuelled nearly 5,000 ships in her role as a jetty without significant incident, one of the last ships to come alongside *Warrior* – the 3,800-ton oil tanker *Wadhurst* – did so at too fast a speed and smashed into her bows, breaking off the beak head and a hawse pipe. Divers recovered both and reported no underwater damage. But this was yet another job to add to the list for the restorers.

When one looked at *Warrior* it was apparent that it was more of a rebuilding task than a preservation project. So much had been stripped out that she was little more than an empty iron shell. As with all projects at their inception, there were differing points of view over how to tackle it. With the amount of rebuilding being greater than restoration there was a suggestion that a shipbuilder

leaking in from above – fortunately she was not leaking below the waterline.

In 1977 the MoD announced that the Llanion oil depot would close and they would have no further use of *Warrior* after April 1978. The MoD sought from the Maritime Trust what their

be contracted to complete the job. Others preferred a more prudent approach with research and preservation taken step by step that would, importantly, control costs.

There was a big advantage that much of *Warrior*'s information and original drawings were preserved in both the Science and Greenwich Maritime Museums. Despite this plethora of data, research would still be required; production drawings would need to be prepared and manufacturing facilities found.

The search began for a site where *Warrior* could be restored and a number of locations in the north of England were considered. Emerging as the best was Hartlepool on the north-east coast. It had the advantage of a berth with good potential for shoreside facilities and most importantly a source of skilled labour. There was high unemployment in the area due to the recent closure of the British Steel works and the docks had been run down. It was estimated that the task would employ about 100 people and would therefore be eligible for Manpower Services Commission assistance.

Warrior was officially handed to the Maritime Trust on 12 August 1979. John Smith's Manifold Trust had agreed to underwrite the project for the estimated rebuilding sum of between £3 and £8 million, on condition that other funding was sought.

At this time the Maritime Trust decided that a subsidiary trust should be created to manage exclusively *Warrior* and this was instigated as the Ship Preservation Trust (SPT). Maldwin Drummond became its chairman. The SPT eventually evolved into the Warrior Preservation Trust that continues to this day.

The London Borough of Newham made another case for homing *Warrior* but Portsmouth stepped in with a promise of building a completely new jetty exclusively for her in Portsmouth Dockyard.

In Hartlepool the Old Coal Dock was *Warrior*'s designated berth but it needed to be dredged to accommodate her. This was among the first significant costs to the SPT, the work exceeding £35,000.

A week after *Warrior* was given to the Maritime Trust she was towed from Milford Haven by the Alexandra Towing Co. – a company directly descended from the towing business of James Watkins who had brought *Warrior* down the Thames in 1861. *Warrior* arrived at Hartlepool on 7 September 1979. Now the work could begin.

BELOW The tug *Eskgarth* bringing *Warrior* **back to life.** (WPT)

Chapter Seven

Restoring *Warrior*

HMS *Warrior* is the largest maritime restoration project ever undertaken and the most complete historic warship afloat in the world today. She has been brought back to life by the dedication of the many people who gave of their skills and their time.

OPPOSITE **HMS *Warrior* afloat in Portsmouth Harbour in 2004.** *(WPT)*

One of the earliest decisions that the Maritime
Trust made was that *Warrior* would be restored
to her 1860 condition – exactly as she was
when she entered service with the Royal Navy.
It was a straightforward decision as *Warrior* had
been at that time the most powerful warship in
the world and nothing had occurred during her
active life that would give grounds for rebuilding
her to a different era.

Regrettably the initial contractor at Hartlepool
chosen for the work proved unsatisfactory
and was discharged. In April 1980 the Ship
Preservation Trust, led by Maldwin Drummond,
took over the management, with Admiral
Patrick Bayly and later Tom Dulake as directors.
The Manifold Trust had to provide the bulk of
the money to keep the project alive. It was
attracting little outside funding, either from the

public or the government, and as a result the
project's aspirations slowed. The Manpower
Services Commission was the only exception –
they provided what support they were able to
within their own limitations.

When work commenced, the project had
nothing in the way of tools, scaffolding, ladders,
gangways and even furniture for the offices.
All of this had to be obtained. Starting with a
small workforce of no more than eight men, the
immediate task was making the ship accessible
and safe to work on. This was a significant job
in itself – the gangways, ladders and guardrails
were fitted and services such as lighting, fire
protection and temporary weather proofing
were installed. Access into the lower decks was
hazardous to say the least.

The Coal Dock berth at Hartlepool had no
facilities ashore but the Old Custom House
100yd away was acquired for offices that would
house the administration staff and drawing
office. A small museum was established,
predominantly to show prospective donors what
was being found. Other temporary workshops
were erected on the quayside.

Having established access on to the ship, the
removal of the detritus of decades began. With
only a small workforce, this task was ongoing for
nearly 2 years. As the ship was slowly cleared,
researchers began to establish what remained
that was original and of the period.

Structurally speaking, the hull was intact
and the restorers took the simplistic view
that if it was iron, it was original. Surprisingly,
areas of the deck planking on the gun deck

LEFT Early days.
(WPT)

BELOW LEFT
**Berthed alongside the
quay that was to be
Warrior's home for
8 years.** *(WPT)*

BELOW **The gun deck
at the beginning of the
restoration.** *(WPT)*

and in particular the lower deck had survived
reasonably intact. The planking on the half
deck, where the captain's cabin is located, was
almost completely sound. The two capstans on
the gun deck were original; though the Brown's
patent steam-driven capstan was not there in
1861, there was no justification in removing a
120-year-old artefact – therefore, it remained.
The galley stove and the galley's brick floor had
also survived. Though the forward magazine
had disappeared, the aft magazine was intact.

A number of artefacts were discovered within
the depths of the ship. A few of the iron-runged
ladders remained, the stove to the drying room
was still in place, an original coal trolley was
found in one of the coal bunkers and a pistol
crocus stowage was discovered in the bilge.
There was little else.

What was not needed? The poop deck
structure was demolished and that led to
the major task of removing the thick layer of

concrete (200 tons of it) that had been laid the
length of the upper deck. The alkaline concrete
had corroded the iron deck beneath, which
resulted in much of it having to be replaced with

BELOW **The cable
deck in 1982.** *(WPT)*

steel plating. This was followed by the need to replace all the bulwark timbers and metal supports and there were concerns that the iron armour and teak backing may also be corroded and rotten, also needing replacing.

The workforce grew with the recruitment of over 25 shipwrights and joiners and apprentices from the Manpower Services Commission joining the project. The project now fell into two distinct roles, one being research, driving the second, the actual restoration itself.

As the debris was cleared, examination of the ship showed a multitude of empty holes, bolts, openings and brackets. What did they all do and what was their purpose? The problem was that ships such as *Warrior* and her contemporaries were in an historical era that had not been studied or researched in any great depth at that time. On the positive side, the Victorians were very good at recording everything they did – drawings, documents and even photographs were scattered around the country either in public archives such as the Maritime Museum at Greenwich, the Public Record Office, the Science Museum or in private collections. *Warrior*'s construction drawings were found at Greenwich. The details were gathered and with such success that the amount of information became almost overwhelming. Much debate began as to what was the definitive data that could be taken forward for the draughtsmen to turn into manufacturing drawings.

Here another dilemma faced the restorers. Though the decision to bring the ship to her original 'as built' condition had been made, it was prohibitive in terms of both cost and skills to try to remanufacture the myriad components and fittings in their original materials and to be functional. There were certainly no plans to take *Warrior* back to sea and certainly no intent to have guns that could be fired. Therefore every alternative had to be explored to achieve the rebuild with an eye firmly on costs, the manufacturing abilities that were available and time.

Surprisingly there was no national inventory of historic guns and there was difficulty in tracking down an actual model of the 68lb cannon, but through the perseverance and dedication of the researchers two original

cannon were located, one in Chatham Dockyard and one in Woolwich Arsenal. The one in Woolwich had lain out in the open among a collection of smaller guns.

The Armstrong 110lb gun was altogether another problem. It had not been a successful weapon in service and on being withdrawn had disappeared. There were a number of versions built and the Royal Armouries in Fort Nelson, Portsmouth, possessed a 72cwt model on display. *Warrior*, however, had the 82cwt version and a surviving example could not be found. William Armstrong and the Army Ordnance department seemed to have done a very good job of losing all information on this problematical weapon! Though drawings and photographs were eventually found, new drawings would have to be produced and the weapons manufactured from scratch.

It was well after the restoration had begun that two 82cwt guns were located. One was in the Victoria Barracks, Sydney, Australia, and the other stood at the base of the war memorial at St Helier, Jersey. From a logistical perspective, the States of Jersey were first approached and through their kind co-operation the gun was made available to enable a copy to be made.

On making enquiries it was discovered that foundry businesses still retained the ability to recast the cannon and to remake the Armstrong guns in iron, but the cost was prohibitive. Therefore a plastics fabrication company, E&F Plastics (conveniently in Hartlepool), were approached and from the two surviving examples they were able to produce moulds from which fibreglass replicas were manufactured. They are indistinguishable from the originals – until you tap them.

The carriages for both the gun and cannon were a different matter and these were manufactured from elm wood in the Preservation Trust's own workshops. These are definitive copies of the originals.

Money and grants were few and far between in coming through and the Manifold Trust was almost solely financing the whole project. In an attempt to alleviate matters, Tony Bridgewater was appointed in May 1981 with the dual task of fund-raising and development officer. Also joining, on a secondment from Cleveland Educational Authority, was Walter Brownlee, an experienced merchant seaman who had turned to teaching history and archaeology. He brought knowledge and practical advice and started the photographic record of the restoration.

The whole of 1981 had been taken up with consolidating the management and the resources, clearing the ship of detritus and tackling the upper deck restoration. Later a pontoon was acquired that enabled the beak head to be repaired after its damage at the Llanion depot. From a Bradford warehouse that was being demolished, 20,000sq ft of pine flooring was recovered, destined for the upper deck.

Early in 1982 came the need to finally decide where *Warrior* was eventually to be berthed. Portsmouth remained the prime candidate, but London and possibly Chatham Dockyard, were options that had to be considered. The Ship Preservation Trust confirmed that Portsmouth offered the best facilities, having not only a deepwater berth but also the adjacent historic dockyard (with HMS *Victory* and the *Mary Rose* arriving that year). *Warrior* was of such size and robust construction that she could stay afloat, which was almost unique among the preserved ships of her age.

Captain John Wells now joined the research team and brought considerable experience and enthusiasm, culminating in him writing the definitive book, *The Immortal Warrior*. He also acted as the liaison officer between the Preservation Trust and Portsmouth City Council. Wells tracked down Arthur Cochrane's papers, including the ship's log books and his diaries of the time when he was captain of *Warrior*; he had kept them on leaving the ship. Arrangements were made for these to be acquired by the Museum of the Royal Navy at Portsmouth. The Museum had also acquired the journal of midshipman Henry Murray that contained an invaluable sketch of each deck of the ship.

The flow of information grew at a prolific rate. Photography was a new technology of the Victorian age and pictures of superb quality were found, not necessarily just of *Warrior*, but of ships of the era that gave further clues and information on particular details.

During 1982 the significant task of rebuilding the bulwarks and the restoration of the upper deck planking was begun. Jack Whitehead

and Norman Gaches were commissioned to reproduce *Warrior*'s figurehead.

From every corner of the country's numerous military stores and museums came information and artefacts: a ship's bell, hammocks of the period that were still in a naval store, 1850 Admiralty compasses. The Tower of London loaned pikes, cutlasses, boarding pikes and tomahawks. Even cooking utensils and 500 metal plates and bowls of the era were discovered.

The year 1983 also brought final confirmation of *Warrior*'s destination with Portsmouth City Council approving the £1.5 million jetty exclusively for the ship just inside the Victory Gate of the dockyard. It would be a prime position where the ship would be clearly seen by everyone arriving in Portsmouth, whether by road, rail or the numerous ferries entering the harbour.

The bulwarks and upper deck work took the project well into 1983 with the deck planking laid in the traditional manner. The seams were filled with hot pitch and oakum, resurrecting old skills among the workforce. Continuing around all this work was the seemingly endless task of removing the layers of paint. Analysis of samples showed that in certain places there were up to 120 different coats in various colours! Health and safety precautions were necessary, as much of the paint was lead-based. The cleaned metalwork was coated with a preservative and then a white paint.

The team now turned to the other jobs that were of seemingly increasing magnitude, the

BELOW The upper deck planking being laid in the 1980s. *(WPT)*

masts, spars and rigging. Detailed drawings were produced and the mizzenmast was the first to be tackled, manufactured from steel piping that had come from the North Sea oil industry rather than timber as the original had been. It was lifted into place in September 1984 and the mainmast followed in November of that year. The foremast and bowsprit were shipped in February 1985. Directly associated with the mast was the rigging, much of the rope for which was made in Chatham Dockyard rope walk. There are many conflicting estimates of the length of the ship's rope, but, whichever is correct, it is measured in miles.

Continuing in parallel with all the other activities was the task of making the ship accessible and friendly for the public with modern amenities having to be installed. These included electrical systems for lighting, public address, fire detection and – by no means least – sanitation.

During late 1983 significant changes to the management of the project took place. The Maritime Trust transferred its interests in the Ship Preservation Trust to the Manifold Trust, with Vice Admiral Bayly resigning. Lord Strathcona and Sir Peter Vanneck joined the SPT board. A Ship Committee was formed that would run the day-to-day management with Tom Dulake as chairman.

The largest empty spaces in *Warrior* were the stoke holds and engine room; the original machinery and boilers had long since been removed. Old Belleville boilers were still in place that had been fitted to drive electrical generators when *Warrior* was *Vernon III* in the 1900s. These were removed, one being donated to the Science Museum.

From the very start it had been the intention to rebuild the engines and boilers with the engines having some functionality to demonstrate better to visitors how they worked. It was a far-reaching ambition and it was some two years into the restoration before serious attention was turned towards achieving this daunting goal.

A separate team was assembled with Jim Wilson appointed as the project engineer in late 1984, bringing to the task his vast experience of propulsion engines. He had served an apprenticeship in shipbuilding as a marine engineer installing steam engines into ships during the Second World War, and then went to sea becoming an engineering officer in the merchant navy. Knowing his way around the shipbuilding world, he quickly recruited two highly competent engineering draughtsmen in Mick Cookman and John Poulton. Along with Jim, they formed the core of the engineering team which, as reconstruction began, drafted in many skilled men from the shipyards that had recently closed in the area.

As had been the case with the guns, alternatives had to be explored in manufacturing the engines and boilers, such that they looked correct but not necessarily from the materials of the originals. Once again researchers were put to the task of finding drawings and information on the engines and boilers. Though *Warrior* drawings were available, they were not sufficiently detailed to be of use. Luckily the National Maritime Museum at Greenwich had a full and detailed set of *Warrior*'s sister ship, *Black Prince*. The Science Museum had a number of models of Penn trunk engines that had been used on many of the early 'Black Battle Fleet' ships.

With this information, the drawing office began the task of replicating Victorian engineering. The use of alternative modern materials and presentation techniques were explored, attributed more to what would be used in the theatrical or film industries. The boilers in the stoke holds, especially, lend themselves more to a theatre set – only the face of the boilers was replicated on to a framework – but it reproduces a wonderful atmospheric impression of the originals.

When it came to the engines, in many cases it was found that returning to cast iron or steel was the most practical solution as the knowledge, skills and experience of the restoration team came from working with these materials. The result is one of the highlights of *Warrior*. A visitor, having passed through the boiler room, emerges up into the engine room where the massive cylinders turn the crankshaft, dramatically showing the sense of scale and power and the engineering genius and ability of the Victorian engineers. Jim Wilson admitted that rebuilding the engines was the highlight of his long career and he said that the team he had assembled were the very best he had ever worked with. He consequently received an award from the Institute of Mechanical Engineers.

There was some discussion at this time of restoring the funnel-lowering gear such that it

LEFT Restoration name plate in the engine room. *(Author)*

BELOW Rebuilding the commander's and paymaster's cabins on the half deck, November 1982. *(WPT)*

functioned but, after due consideration (again with cost and time being factored in), the idea was dropped.

By the end of 1985 many of the guns and

LEFT *Warrior* at the Hartlepool berth. *(WPT)*

mess tables had begun to appear on the
gun deck, cabins were being completed on
the half deck and the boiler and engine room
were all but finished. *Warrior* was coming
back to life. From all directions artefacts and
equipment continued to be sourced: clay
pipes, photographs of the period, galley pots
and pans (many coming from the royal yacht
Britannia) and Royal Worcester Spode providing
wardroom plates in an 1862 design with
naval crown patterns. In addition, RN College

Dartmouth donated silverware, uniforms for the
crew were remade and tailors Gieves & Hawkes
made a frock coat for the appointed captain.

A captain had indeed been appointed
to *Warrior*. Captain Colin Allen would take
command of the ship on his retirement from
the Royal Navy. His immediate task was putting
together the team for when the ship arrived in
Portsmouth and opened to the public.

It had been planned that *Warrior* would
come to Portsmouth in September 1986
but in January of that year an assessment of
outstanding work placed that departure date
in some doubt unless there was a significant
increase in effort. The Manifold Trust advised
that they would not be able to increase the
financing of the project that was already costing
them £90,000 per month.

A new date of spring 1987 was agreed upon
that proved to be advantageous to all parties
as it gave Portsmouth time to complete the
new jetty and coincided with the start of the
tourist season. In Hartlepool it provided time
to fully complete the restoration and gave the
workforce six months' further employment.

Prior to her departure for Portsmouth, the
Duke of Edinburgh visited the ship and the
restoration team. In many respects it was he
and John Smith who had started the whole

LEFT HMS *Trincomalee* **ready to go to Hartlepool as** *Warrior* **was on her way to Portsmouth.** *(NMRN)*

venture off and it must have been satisfying for them to see the result of their efforts. In June 1987 *Warrior* left Hartlepool for the 350-mile voyage to Portsmouth. She had cost £8 million to restore. John Smith received a well-deserved knighthood in 1988.

It was a sad day in Hartlepool when *Warrior* left, especially for all those who had been employed by the project and had given so much effort in bringing this fantastic ship back to life. But there was a postscript: as *Warrior*

left Hartlepool, discussion was under way to reverse the process and send a Portsmouth ship to Hartlepool for restoration and display. In Portsmouth Harbour was the training ship TS *Foudroyant*, an ex-Royal Navy frigate built from teak in Bombay in 1817 as HMS *Trincomalee*. She became part of the National Historic Fleet under her original name and went to be fully rebuilt at Hartlepool, where she remains to this day open to the public as the oldest surviving warship that is still afloat.

LEFT Like *Warrior*, the restored HMS *Trincomalee* remains afloat. *(NMRN)*

How many *Warriors*?

There have been six HMS *Warriors* in Royal Navy service. The first was a wooden 74-gun third-rate ship of the line built in 1781 at Portsmouth. She had an eventful career: almost immediately on entering service she took part in the Battle of the Saintes in the West Indies in 1782 where a combined French and Spanish fleet were defeated, and then in 1801 was present at the Battle of Copenhagen but did not engage the enemy as she was in the Reserve Squadron. In 1805 she participated in the Battle of Cape Finisterre, being the only ship to have no casualties although she was described after the battle as being 'much torn'.

ABOVE A superb diorama of the launch of the first *Warrior*, 18 November 1781. *(NMRN)*

RIGHT The surviving 1860 *Warrior* at her berth in Portsmouth. *(Author)*

With the end of the Napoleonic Wars, the navy was reduced in numbers and she became a receiving ship in 1818 where newly recruited sailors were accommodated before being assigned to a ship. After 1840 she was designated a prison ship and was eventually broken up in 1857.

The second *Warrior* is the preserved ship that this book focuses on.

The third Warrior was a Duke of Edinburgh-class armoured cruiser launched in 1905. She was badly damaged and disabled at the Battle of Jutland on 31 May 1916 and foundered a day later. Fortunately the majority of her crew were saved.

His Majesty's Patrol Yacht *Warrior* was a yacht requisitioned by the Royal Navy in 1917. She was originally built at Troon in Scotland as a 1,266-ton luxury yacht for the American railway tycoon Fredrick William Vanderbilt. She deployed to the West Indies Station where she made a number of visits to the United States

and was sold in 1920, reverting to a yacht under various names and owners. She then regained her name (with a suffix) of *Warrior II* in 1937 when purchased by Sir Hugh Cunliffe-Owen, only to be requisitioned again at the commencement of the Second World War and fitted with 12lb guns. On 11 July 1940 she was bombed by German Luftwaffe aircraft and sunk in the English Channel off Portland, where she remains to this day as a popular wreck for divers.

The fifth HMS *Warrior* was a Colossus-class light aircraft carrier built by Harland and Wolff in Belfast between 1942 and 1945. She was transferred to the Canadian Navy in 1946 as the HMCS *Warrior* and returned to the Royal Navy in 1948 where she remained in service until 1958, subsequently being transferred to the Argentinian Navy and renamed ARA *Independencia*. She survived until 1971 when she was withdrawn and scrapped.

BELOW The 1905 light cruiser *Warrior* **was lost at the Battle of Jutland in 1916.** *(NMRN)*

In the 1960s new NATO Headquarters were established at RAF Northwood, Middlesex, where the Commander-in-Chief Home Fleet was to be based. His headquarters took the name *Warrior*.

In 1996 it became part of the Permanent Joint Headquarters (PJHQ) and the *Warrior* name rescinded, so HMS *Warrior* 1860 once again became the only ship to bear the name.

Index

Sources

Inkersole, Royston and Wilson, Alastair, *Raising Steam* (Private publication, Warrior Preservation Trust)
Lambert, Andrew, *Warrior – Restoring the World's First Ironclad* (Conway Maritime Press, 1987)
Wells, Captain John, *The Immortal Warrior* (Kenneth Mason Publications, 1997)